ISSN 0044-2267

ZEITSCHRIFT FÜR ANGEWANDTE MATHEMATIK UND MECHANIK

INGENIEURWISSENSCHAFTLICHE FORSCHUNGSARBEITEN

BAND 62   1982   HEFT 10

AUS DEM INHALT

AKADEMIE-VERLAG · BERLIN

ISSN 0044-2267   Z. angew. Math. u. Mech., Berlin 62 (1982) 10, 501—572   EVP 18,— M   34115

INHALT

Hauptaufsätze

Kleine Mitteilungen

Wir bitten, **Manuskriptsendungen** zweifach (Original und eine Kopie, sprachlich einwandfrei, Formeln mit Maschine oder in Druckschrift geschrieben) an folgende Anschrift zu richten:

Zeitschrift für Angewandte Mathematik und Mechanik, z. Hd. Herrn Prof. Dr. Günter Schmidt
Institut für Mechanik der Akademie der Wissenschaften der DDR
DDR-1199 Berlin, Rudower Chaussee 5

Zu den Arbeiten, die als Hauptaufsätze bestimmt sind, ist auf gesondertem Blatt eine Zusammenfassung von 5 bis 10 Zeilen in englischer und (möglichst) deutscher und russischer Sprache beizufügen. Ausführliche Hinweise für die Autoren, um deren strikte Berücksichtigung gebeten wird, finden sich im Anschluß an das Inhaltsverzeichnis des letzten Jahrganges.

Die Autoren erhalten von den Hauptaufsätzen 75, von den Kleinen Mitteilungen 25 Sonderdrucke ohne Berechnung, darüber hinaus weitere Sonderdrucke gegen Berechnung.

Der Verlag behält sich für alle Beiträge das Recht der Vervielfältigung und Übersetzung vor.

Bestellungen sind zu richten

- **in der DDR** an den Postzeitungsvertrieb unter Angabe der Kundennummer des Bestellenden oder an den AKADEMIE-VERLAG, DDR-1086 Berlin, Leipziger Str. 3—4, PSF 1233
- **im sozialistischen Ausland** an eine Buchhandlung für fremdsprachige Literatur oder an den zuständigen Postzeitungsvertrieb
- **in der BRD und Berlin(West)** an eine Buchhandlung oder an die Auslieferungsstelle KUNST UND WISSEN, Erich Bieber OHG, D-7000 Stuttgart 1, Wilhelmstraße 4—6
- **in den übrigen westeuropäischen Ländern** an eine Buchhandlung oder an die Auslieferungsstelle KUNST UND WISSEN, Erich Bieber GmbH, CH-8008 Zürich, Dufourstraße 51
- **im übrigen Ausland** an den Internationalen Buch- und Zeitschriftenhandel; den Buchexport, Volkseigener Außenhandelsbetrieb der Deutschen Demokratischen Republik, DDR-7010 Leipzig, Postfach 160; oder an den AKADEMIE-VERLAG, DDR-1086 Berlin, Leipziger Straße 3—4, PSF 1233

ZEITSCHRIFT FÜR ANGEWANDTE MATHEMATIK UND MECHANIK

Herausgeber und Chefredakteur: Prof. Dr. Günter Schmidt. Redaktion: Dr. Winfried Heinrich, Dr. Horst Weinert, Dipl.-Math. Friedhild Dudel, Bernhilde Schwertfeger, Institut für Mechanik der Akademie der Wissenschaften der DDR. Verlag: Akademie-Verlag, DDR-1086 Berlin, Leipziger Straße 3—4; Fernruf: 2236221 oder 2236229; Telex-Nr.: 114420; Bank: Staatsbank der DDR, Berlin, Kto.-Nr.: 6836-26-20712. Anschrift der Redaktion: Institut für Mechanik der Akademie der Wissenschaften, DDR-1199 Berlin, Rudower Chaussee 5; Fernruf: 6743639 oder 6743643.

Veröffentlicht unter der Lizenznummer 1282 des Presseamtes beim Vorsitzenden des Ministerrates der Deutschen Demokratischen Republik. Gesamtherstellung: VEB Druckerei „Thomas Müntzer", 5820 Bad Langensalza. Erscheinungsweise: Die Zeitschrift für Angewandte Mathematik und Mechanik erscheint monatlich. Die 12 Hefte eines Jahres einschließlich Tagungshefte bilden einen Band. Bezugspreis je Band 360,— M zuzüglich Versandspesen (Preis für die DDR 216,— M). Bezugspreis je Heft 30,— M (Preis für die DDR 18,— M). Bestellnummer dieses Heftes: 1009/62/10.

AN (EDV) 35937

ZAMM 62, 501 – 509 (1982)

H. Müller

Schwingungsinduzierte Durchflußänderung im Ringraum zwischen konzentrischen Rohren

Im Ringraum zwischen konzentrischen Rohren fließe inkompressible newtonsche Flüssigkeit in axialer Richtung. Durch harmonische Querschwingungen des Innenzylinders wird eine Veränderung des nunmehr zeitlich gemittelten Volumenstromes bei gleichbleibendem Druckgradienten bewirkt, deren Berechnung im Rahmen einer Störungstheorie Ziel dieser Arbeit ist. Es zeigt sich, daß im Fall langsamer Schwingungen Volumenstromerhöhung, im Fall schneller Schwingungen Volumenstromminderung eintritt. Im Bereich mittlerer Frequenzen findet ein monotoner Übergang zwischen diesen Extremzuständen statt. Die Effekte sind um so stärker, je enger der Spalt zwischen den Rohren ist, d. h. je näher deren Radienverhältnis bei 1 liegt.

An incompressible Newtonian fluid filling the gap between two concentric circular pipes flows in axial direction. Harmonic vibrations of the inner pipe in lateral direction affect the fully developed flow and change the mean flow volume for a given value of the axial pressure gradient. The aim of the paper is the calculation of that change of flow volume. Slow vibrations increase the flow volume, rapid vibrations cause a decrease. For intermediate values of the frequency the flow volume changes monotonically between the extreme values for small and for large frequencies. These effects grow with decreasing gap width, i.e. they become more prominent when the radius ratio of the two pipes approaches the value 1.

В кольцевом пространстве между концентрическими трубами течёт несжимаемая ньютоновская жидкость в осевом направлении. Гармонические поперечные колебания внутреннего цилиндра вызывают изменение усреднённого по времени объёмного потока при неизменном градиенте давления. Целью данной работы является определение изменения этого объёмного потока. Медленные колебания вызывают увеличение, а более быстрые колебания — уменьшение объёмного потока. В области средних частот происходит монотонный переход между этими экстремальными состояниями. Чем уже пространство между трубами, т.е. чем ближе отношение их радиусов к единице, тем сильнее проявляются эти эффекты.

1. Einleitung

Strömungen im Ringraum zwischen konzentrischen Rohren treten in der Technik häufig auf. Überall da, wo lange dünne Stäbe axial umströmt werden (das geschieht in der Regel, um sie zu kühlen) kann es infolge von aeroelastischer Instabilität zu Querschwingungen des Stabes kommen. Ein Beispiel sind Brennstabschwingungen in Kernreaktoren. Sie können sowohl selbsterregt sein, als auch von außen erzeugt werden. In der Regel sind sie unerwünscht, da sie zu erheblichen Durchflußänderungen führen können und der Wärmeaustausch stark beeinflußt werden kann.

Eine vollständige mathematische Beschreibung dieses Phänomens scheitert an dessen Kompliziertheit. Abgesehen vom nicht-linearen Charakter der Navier-Stokes-Gleichungen handelt es sich um ein Wechselwirkungsproblem zwischen Strömung und Schwingung, wobei die Bewegung des inneren Zylinders a priori nicht bekannt ist, sondern mitermittelt werden muß. Deshalb wird in dieser Arbeit angenommen, die Bewegung des inneren Zylinders sei dem System in Form einer harmonischen Schwingung von außen fest aufgeprägt, so daß nur die Strömung ermittelt werden muß.

Ziel ist die Aufstellung eines Zusammenhangs zwischen dem zeitlich gemittelten Volumenstrom und Amplitude und Frequenz der Schwingung.

Zu diesem Themenkreis existiert bereits die Arbeit von Farah [4], die für kleine Amplituden die Fälle sehr langsamer und sehr schneller Schwingungen behandelt. Der Fall langsamer Schwingungen beruht auf einer regulären Störungstheorie, die in der vorliegenden Arbeit nicht dargestellt wird. Statt dessen wird eine zu der Farahschen Methode schneller Schwingungen alternative Theorie vorgestellt, die auf einer singulären Störungsrechnung beruht. Veranlaßt durch die Tatsache, daß diese Theorien zu verschiedenen Resultaten führen, wurde Farahs Methode geprüft und festgestellt, daß sie fehlerhaft ist.

2. Die Grundgleichungen

Gemäß Fig. 1 sei r_i der Radius des Innenrohres, r_a der des Außenrohres. Die newtonsche Flüssigkeit (Dichte: ϱ, dyn. Zähigkeit: η, kin. Zähigkeit: $\eta/\varrho = \nu$) fließe aufgrund eines negativen Druckgradienten in z-Richtung. Der Innenzylinder schwingt harmonisch um die konzentrische Ruhelage nach dem Gesetz

$$y = A \cdot \sin \omega t \,. \tag{2.1}$$

Es werden folgende vereinfachende Annahmen gemacht:

(i) Die Strömung ist laminar. Effekte, die von Turbulenz oder dem Ablösen von Grenzschichten verursacht werden, bleiben unberücksichtigt.

(ii) Die Strömung ist ausgebildet. Für die Geschwindigkeit gilt:

$$\frac{\partial \boldsymbol{v}}{\partial z} = 0 \,. \tag{2.2}$$

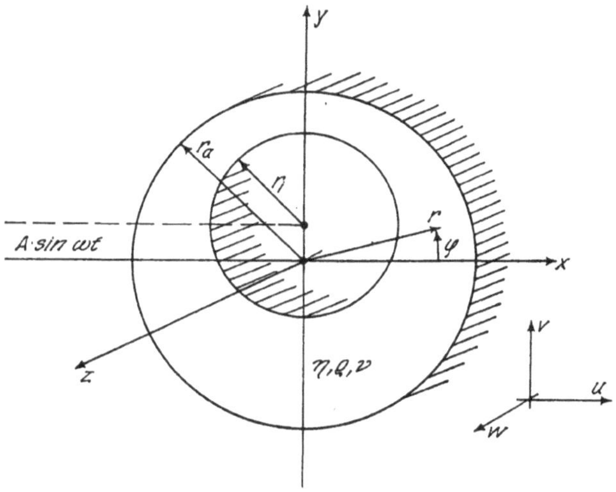

Fig. 1

Für den Druck folgt daher:

$$p(r, \varphi, z, t) = \tilde{p}(r, \varphi, t) - \lambda z \,, \tag{2.3}$$

mit dem negativen Druckgradienten:

$$\lambda = -\frac{\partial p}{\partial z} = \text{const.} > 0 \,. \tag{2.4}$$

(iii) Die Schwingungsamplitude A sei klein gegenüber dem Innenradius r_i. Die Nichtlinearität der NAVIER-STOKES-Gleichungen ist nur zu umgehen durch Entwicklung nach dem Parameter

$$\varepsilon := \frac{A}{r_i} \ll 1 \,. \tag{2.5}$$

Da der Durchfluß nur eine gerade Funktion der Schwingungsamplitude sein kann, muß die Störungstheorie von zweiter Ordnung sein.

An späterer Stelle wird sich zeigen, daß man nur durch weitere vereinfachende Annahmen zu einer analytischen Lösung gelangen kann.

2.1 Das Geschwindigkeitsfeld

Aufgrund der Annahme (ii) entfallen die Elemente $\frac{\partial u}{\partial z}, \frac{\partial v}{\partial z}, \frac{\partial w}{\partial z}$ des Geschwindigkeitsgradienten, so daß die Längsströmung ohne Einfluß auf die Querströmung ist (nicht aber umgekehrt). Da somit die Querströmung ein rein ebenes Problem darstellt, für das sich eine Stromfunktion Ψ einführen läßt, ist folgende Aufspaltung des Geschwindigkeitsvektors zweckmäßig:

$$\boldsymbol{v} = -\boldsymbol{e}_z \times \operatorname{grad} \Psi + w\boldsymbol{e}_z \,. \tag{2.6}$$

Die Stromfunktion Ψ hat die Wirbeltransportgleichung zu erfüllen — hier in Polarkoordinaten —:

$$\left(\frac{\partial}{\partial t} - \nu\Delta\right)\Delta\Psi = \frac{1}{r}\left(\Psi_r \frac{\partial}{\partial \varphi} - \Psi_\varphi \frac{\partial}{\partial r}\right)\Delta\Psi \,. \tag{2.7}$$

Der Δ-Operator bezieht sich natürlich nur auf die Querschnittsfläche.

Die Geschwindigkeitskomponente in Längsrichtung w genügt der dritten Komponente der NAVIER-STOKES-Gleichung:

$$\left(\frac{\partial}{\partial t} - \nu\Delta\right)w = \frac{\lambda}{\varrho} + \frac{1}{r}\left(\Psi_r \frac{\partial}{\partial \varphi} - \Psi_\varphi \frac{\partial}{\partial r}\right)w \,. \tag{2.8}$$

2.2 Die Randbedingungen

Die Randbedingungen sind die des Haftens:

$$\boldsymbol{v} = 0: \quad \text{am Außenrand:} \quad r = r_a \,, \qquad \boldsymbol{v} = A\omega \cos \omega t \cdot \boldsymbol{e}_y: \quad \text{am Innenrand} \,. \tag{2.9), (2.10}$$

In Komponenten lauten sie:

$$\Psi_r = \Psi_\varphi = w = 0: \quad r = r_a \,, \tag{2.11}$$

$$\left.\begin{aligned} \Psi_r &= -A\omega \cos \omega t \cos \varphi \,, \\ \Psi_\varphi &= A\omega r \cos \omega t \sin \varphi \,, \\ w &= 0 \end{aligned}\right\} \text{am Innenrand} \,. \tag{2.12}$$

Die Gleichung des Innenrandes ist die eines Kreises:

$$x^2 + (y - A \sin \omega t)^2 = r_i^2 . \tag{2.13}$$

In Polarkoordinaten lautet sie:

$$r^2 - 2Ar \sin \omega t \sin \varphi + A^2 \sin^2 \omega t = r_i^2 . \tag{2.14}$$

2.3 Der Durchfluß

Der zeitlich gemittelte Durchfluß ist definiert als:

$$\dot{V} = \frac{1}{T} \int_0^T \iint_{\mathcal{B}(t)} wr \, dr \, d\varphi \, dt , \tag{2.15}$$

mit der zeitabhängigen Ringraumfläche $\mathcal{B}(t)$ und der Schwingungsdauer:

$$T = 2\pi/\omega . \tag{2.16}$$

2.4 Entdimensionierung

Um das Problem leichter entwickeln zu können, wird folgende Entdimensionierung durchgeführt:

$$\frac{r}{r_i} =: r^*; \qquad \omega t =: t^*; \qquad \Psi/A\omega r_i =: \Psi^*; \qquad w\Big/\frac{\lambda r_i^2}{\eta} =: w^*; \qquad \frac{\Delta}{r_i^2} \equiv : \Delta^*; \tag{2.17}$$

$\mathcal{B}^*$: $\mathcal{B}$ im dimensionslosen r^*, φ, t^*-Raum.
Die Entdimensionierung der Gleichungen (2.7), (2.8), (2.11), (2.12), (2.14) und (2.15) liefert unter Weglassung der Sterne:

$$\left(\Omega \frac{\partial}{\partial t} - \Delta\right) \Delta \Psi = \frac{\varepsilon \Omega}{r} \left(\Psi_r \frac{\partial}{\partial \varphi} - \Psi_\varphi \frac{\partial}{\partial r}\right) \Delta \Psi , \tag{2.18}$$

$$\left(\Omega \frac{\partial}{\partial t} - \Delta\right) w = 1 + \frac{\varepsilon \Omega}{r} \left(\Psi_r \frac{\partial}{\partial \varphi} - \Psi_\varphi \frac{\partial}{\partial r}\right) w , \tag{2.19}$$

$$\Psi_r = \Psi_\varphi = w = 0 \quad \text{für} \quad r = R , \tag{2.20}$$

$$\Psi_r = -\cos t \cos \varphi , \qquad \Psi_\varphi = r \cos t \sin \varphi , \qquad w = 0 \quad \text{auf dem Innenrand,} \tag{2.21}$$

der gegeben ist durch

$$r^2 - 2\varepsilon r \sin t \sin \varphi + \varepsilon^2 \sin^2 t = 1 , \tag{2.22}$$

$$\dot{V} = \frac{\lambda r_i^4}{2\pi\eta} \int_0^{2\pi} \iint_{\mathcal{B}(t)} wr \, dr \, d\varphi \, dt . \tag{2.23}$$

Die entstandenen Parameter haben folgende Bedeutung:

$$\text{Radienverhältnis:} \quad \frac{r_a}{r_i} = R; \tag{2.24}$$

$$\text{Frequenz:} \quad \frac{\omega r_i^2}{\nu} = \Omega; \tag{2.25}$$

$$\text{Amplitude:} \quad \frac{A}{r_i} = \varepsilon . \tag{2.26}$$

Aufgrund seiner Nichtlinearität ist das System (2.18)—(2.23) nicht geschlossen lösbar. Es muß daher von Annahme (iii) Gebrauch gemacht und eine Störungstheorie zweiter Ordnung bezüglich ε durchgeführt werden.

3. Störungstheorie für kleine Amplituden

Mit Hilfe des Störungsansatzes

$$\Psi = \Psi^{(0)} + \varepsilon \Psi^{(1)} + \dots , \qquad w = w^{(0)} + \varepsilon w^{(1)} + \varepsilon^2 w^{(2)} + \dots \tag{3.1), (3.2}$$

soll das System (2.18)—(2.23) entwickelt werden. Da in (2.19) die Stromfunktion nur in der Form $\varepsilon \cdot \Psi$ auftritt, ist in (3.1) ein Term $\varepsilon^2 \cdot \Psi^{(2)}$ überflüssig.

3.1 Das Geschwindigkeitsfeld

Die Ansätze (3.1), (3.2) angewendet auf (2.18), (2.19) liefern durch Abgleich der Potenzen von ε:

$$\left(\Omega\frac{\partial}{\partial t}-\Delta\right)\Delta\Psi^{(0)}=0\,, \tag{3.3}$$

$$\left(\Omega\frac{\partial}{\partial t}-\Delta\right)\Delta\Psi^{(1)}=\frac{\Omega}{r}\left(\Psi_r^{(0)}\frac{\partial}{\partial\varphi}-\Psi_\varphi^{(0)}\frac{\partial}{\partial r}\right)\Delta\Psi^{(0)}\,, \tag{3.4}$$

$$\left(\Omega\frac{\partial}{\partial t}-\Delta\right)w^{(1)}=\frac{\Omega}{r}\left(\Psi_r^{(0)}\frac{\partial}{\partial\varphi}-\Psi_\varphi^{(0)}\frac{\partial}{\partial r}\right)w^{(0)}\,, \tag{3.5}$$

$$\left(\Omega\frac{\partial}{\partial t}-\Delta\right)w^{(2)}=\frac{\Omega}{r}\left(\Psi_r^{(0)}\frac{\partial}{\partial\varphi}-\Psi_\varphi^{(0)}\frac{\partial}{\partial r}\right)w^{(1)}+\frac{\Omega}{r}\left(\Psi_r^{(1)}\frac{\partial}{\partial\varphi}-\Psi_\varphi^{(1)}\frac{\partial}{\partial r}\right)w^{(0)}\,. \tag{3.6}$$

Der Anteil $w^{(0)}$ ist die stationäre, rotationssymmetrische Grundströmung; sie ist bekanntlich durch folgenden Ausdruck gegeben:

$$w^{(0)}=-\frac{1}{4}(r^2-1)+\frac{1}{4}(R^2-1)\frac{\ln r}{\ln R}\,. \tag{3.7}$$

3.2 Die Randbedingungen

Die Randbedingungen am Außenrand (2.20) gehen über in:

$$\Psi_r^{(0)}=\Psi_\varphi^{(0)}=\Psi_r^{(1)}=\Psi_\varphi^{(1)}=w^{(1)}=w^{(2)}=0\quad\text{für}\quad r=R\,. \tag{3.8}$$

Die entwickelte Innenrandgleichung (2.22) lautet:

$$r=1+\varepsilon\sin t\sin\varphi-\frac{\varepsilon^2}{2}\sin^2 t\cos^2\varphi+\dots\,. \tag{3.9}$$

Die Entwicklung der inneren Randbedingungen (2.21) ist etwas schwieriger, denn es muß beachtet werden, daß die Funktionen sowohl direkt als auch über ihr r-Argument von ε abhängen. Dies sei am Beispiel der Randbed. (2.21) für Ψ_r erläutert:

$$-\cos t\cos\varphi=\Psi_r\left(1+\varepsilon\sin t\sin\varphi-\frac{\varepsilon^2}{2}\sin^2 t\cos^2\varphi,\varphi,t,\varepsilon\right)=$$
$$=\Psi_r^{(0)}(1+\varepsilon\dots)+\varepsilon\Psi_r^{(1)}(1+\varepsilon\dots)+\dots=\Psi_r^{(0)}(1)+\varepsilon\Psi_{rr}^{(0)}(1)+\varepsilon\Psi_r^{(1)}(1)+\dots\,. \tag{3.10}$$

Die auf den festen Rand $r=1$ projizierten Randbedingungen ergeben sich somit zu:

$$\Psi_r^{(0)}=-\cos t\cos\varphi\,,\qquad \Psi_r^{(1)}=-\sin t\sin\varphi\cdot\Psi_{rr}^{(0)}\,,$$
$$\Psi_\varphi^{(0)}=\cos t\sin\varphi\,,\qquad \Psi_\varphi^{(1)}=-\sin t\sin\varphi\cdot\Psi_{r\varphi}^{(0)}+\sin t\cos t\sin^2\varphi=0\,,$$
$$w^{(1)}=-\sin t\sin\varphi\cdot w_r^{(0)}=-\sin t\sin\varphi\left(-\frac{1}{2}+\frac{R^2-1}{4\ln R}\right), \tag{3.11}$$
$$w^{(2)}=-\sin t\sin\varphi\cdot w_r^{(1)}-\tfrac{1}{2}(\sin^2 t\sin^2\varphi\cdot w_{rr}^{(0)}-\sin^2 t\cos^2\varphi\cdot w_r^{(0)})=$$
$$=-\sin t\sin\varphi\cdot w_r^{(1)}+\frac{1}{2}\sin^2 t\sin^2\varphi\left(\frac{1}{2}+\frac{R^2-1}{4\ln R}\right)+\frac{1}{2}\sin^2 t\cos^2\varphi\left(-\frac{1}{2}+\frac{R^2-1}{4\ln R}\right).$$

3.3 Der Durchfluß

Bei der Entwicklung des Durchflusses (2.23) ist darauf zu achten, daß sowohl der Integrand als auch die Integrationsgrenzen von ε abhängen:

$$\dot V=\frac{\lambda r_i^4}{2\pi\eta}\int_0^{2\pi}\!\!\int_0^{2\pi}\left\{\int_{1+\varepsilon\sin t\sin\varphi-\frac{\varepsilon^2}{2}\sin^2 t\cos^2\varphi}^{R}(w^{(0)}+\varepsilon w^{(1)}+\varepsilon^2 w^{(2)})\,r\,\mathrm{d}r\right\}\mathrm{d}\varphi\,\mathrm{d}t=\dot V_0+\frac{\varepsilon^2}{2}\left.\frac{\mathrm{d}^2\dot V}{\mathrm{d}\varepsilon^2}\right|_{\varepsilon=0}+\dots \tag{3.12}$$

Daher muß bei der Differentiation die LEIBNIZ-Regel benutzt werden. Die auftretenden Randterme vereinfachen sich aufgrund der Randbedingungen (3.11). Mit dem bekannten Volumenstrom bei konzentrisch ruhendem Innenzylinder

$$\dot V_0=\frac{\pi\lambda r_i^4}{8\eta}\left(R^4-1-\frac{(R^2-1)^2}{\ln R}\right) \tag{3.13}$$

folgt:

$$\frac{\dot V - \dot V_0}{\dot V_0 \varepsilon^2} = \frac{1}{\pi^2\left(R^4 - 1 - \dfrac{(R^2-1)^2}{\ln R}\right)} \left\{ \int_0^{2\pi}\int_0^{2\pi}\int_1^R w^{(2)} r \,\mathrm{d}r\,\mathrm{d}\varphi\,\mathrm{d}t - \frac{\pi^2}{2}\left(\frac{1}{2} - \frac{R^2-1}{4\ln R}\right)\right\}. \tag{3.14}$$

3.4 Ansätze für die Lösungsanteile

Die Inhomogenitäten (3.8), (3.11) der Dgln. (3.3)—(3.6) legen folgende Ansätze nahe:

$$\begin{aligned} &\Psi^{(0)}(r,\varphi,t) = \cos\varphi \cdot \mathrm{Re}\left(F(r)\,\mathrm{e}^{-it}\right), && \Psi^{(1)}(r,\varphi,t) = \sin 2\varphi \cdot \Phi^{(1)}(r,t),\\ &w^{(1)}(r,\varphi,t) = \sin\varphi\,\mathrm{Re}\left(G(r)\,\mathrm{e}^{-it}\right), && w^{(2)}(r,\varphi,t) = S(r) + \mathrm{per}\,(r,\varphi,t). \end{aligned} \tag{3.15}$$

Der Term per (r, φ, t) deutet an, daß zu S noch Anteile hinzukommen, die entweder periodisch bzgl. der Zeit, oder periodisch bzgl. des Winkels, oder bzgl. beiden sind. Aufgrund der zeitlichen Mittelung und Winkelintegration sind sie daher für das Volumenstromintegral (3.14) ohne Belang und werden nicht weiter betrachtet.

Die Querströmung 1. Ordnung, $\Psi^{(1)}$, stellt sich als proportional zu $\sin 2\varphi$ heraus und bewirkt wegen (3.6) und der Rotationssymmetrie der Grundströmung (3.7) lediglich einen bzgl. des Winkels periodischen Anteil an per (r, φ, t). Daher ist auch sie zur Ermittlung des Volumenstromes überflüssig.

Zur Vereinfachung wird folgender EULERscher Differentialoperator eingeführt:

$$\mathscr{E} :\equiv \frac{\mathrm{d}^2}{\mathrm{d}r^2} + \frac{1}{r}\cdot\frac{\mathrm{d}}{\mathrm{d}r} - \frac{1}{r^2}. \tag{3.16}$$

Für die für den Volumenstrom relevanten Lösungsanteile folgt aus (3.3), (3.5), (3.6), (3.8), (3.11), (3.14):

$$(i\Omega + \mathscr{E})\,\mathscr{E}F = 0 \quad \text{mit} \quad F(R) = F'(R) = 0 \quad \text{und} \quad F(1) = F'(1) = -1, \tag{3.17}$$

$$(i\Omega + \mathscr{E})\,G = \frac{\Omega}{r} F\left(\frac{r}{2} - \frac{R^2-1}{4r\ln R}\right) \quad \text{mit} \quad G(R) = 0 \quad \text{und} \quad G(1) = i\left(\frac{1}{2} - \frac{R^2-1}{4\ln R}\right), \tag{3.18}$$

$$S'' + \frac{1}{r}S' = -\frac{\Omega}{4r}\mathrm{Re}\,(\overline{F}G)' \quad \text{mit} \quad S(R) = 0 \quad \text{und} \quad S(1) = \frac{1}{16}\frac{R^2-1}{\ln R} - \frac{1}{4}\mathrm{Im}\,G'(1), \tag{3.19}$$

$$\frac{\dot V - \dot V_0}{\dot V_0 \varepsilon^2} = \frac{1}{R^4 - 1 - \dfrac{(R^2-1)^2}{\ln R}} \left\{16\int_1^R S r\,\mathrm{d}r - 1 + \frac{R^2-1}{2\ln R}\right\}. \tag{3.20}$$

Die Dgl. (3.19) für S ist von leicht integrierbarer Gestalt. Ihre Lösung, in das Durchflußintegral (3.20) eingesetzt, liefert mit einer partiellen Integration:

$$\begin{aligned} \frac{\dot V - \dot V_0}{\dot V_0 \varepsilon^2} = \frac{1}{R^4 - 1 - \dfrac{(R^2-1)^2}{\ln R}} &\left\{2\Omega\int_1^R \mathrm{Re}\,\overline{F}G\,r\,\mathrm{d}r - \Omega\frac{R^2-1}{\ln R}\int_1^R \mathrm{Re}\,\overline{F}G\frac{\mathrm{d}r}{r} + \right.\\ &\left. + 2\left(1 - \frac{R^2-1}{2\ln R}\right)\mathrm{Im}\,G'(1) - 1 + \left(\frac{R^2-1}{2\ln R}\right)^2\right\}. \end{aligned} \tag{3.21}$$

Die Differentialgleichung für F (3.17) besitzt eine Lösung der Form

$$F(r) = Ar + \frac{B}{r} + CI_1\left(\sqrt{-i\Omega}\,r\right) + DK_1\left(\sqrt{-i\Omega}\,r\right) \tag{3.22}$$

mit den modifizierten BESSEL-Funktionen I_1 und K_1 erster Ordnung. Schwierigkeiten ergeben sich jedoch bei der Suche nach Partikularlösungen für G aus (3.18), da spezielle transzendente Funktionen auftreten, die nicht tabelliert sind. Noch schlimmer wird dieser Sachverhalt bei dem Versuch, Stammfunktionen für die Produktbildungen $\mathrm{Re}\,\overline{F}G \cdot r$ und $\mathrm{Re}\,\overline{F}G/r$ zu finden.

Die Idee, eine komplexe Veränderliche einzuführen, um so die Vorzüge der Funktionentheorie zu genießen, schlägt fehl, da die Funktion $\overline{F}G$ nicht holomorph ist und die Integrale daher wegabhängig sind. Deshalb gibt es nur die beiden Alternativen, entweder das System (3.17), (3.18), (3.21) numerisch zu lösen, was in dieser Arbeit allerdings nicht geschehen soll, oder mittels weiterer Annahmen dieses Systems einer geschlossenen Lösung zugänglich zu machen.

FARAH hat in seiner Arbeit [4] die Annahme sehr langsamer Schwingungen dazu benutzt, eine reguläre Störungsrechnung für kleine Ω durchzuführen. Die Ergebnisse dieser Theorie wurden an einem Beispiel geprüft und als richtig befunden. Daher werden sie hier nicht wiederholt. Anders im Fall schneller Schwingungen: Hier benutzt FARAH im wesentlichen die Gleichungen

$$I_1(z) = \frac{1}{\sqrt{2\pi z}}\mathrm{e}^z\left(1 + O\left(\frac{1}{z}\right)\right), \qquad K_1(z) = \sqrt{\pi/2z}\,\mathrm{e}^{-z}\left(1 + O\left(\frac{1}{z}\right)\right), \tag{3.23}$$

um die auftretenden modifizierten Bessel-Funktionen durch Exponentialfunktionen zu ersetzen. Da es sich zeigt, daß die Approximationen (3.23) nicht einmal zum richtigen Auffinden der asymptotischen Werte für $\Omega \to \infty$ ausreichend sind, sind die Farahschen Ergebnisse schneller Schwingungen falsch.

Im folgenden Kapitel wird daher eine Theorie schneller Schwingungen betrieben, die von Farahs Methode abweicht und zu wesentlich einfacheren und übersichtlicheren Ergebnissen führt. Weil das Problem (3.17), (3.18), (3.21) für $\Omega \to \infty$ in ein singuläres Störungsproblem übergeht, leistet dies eine Grenzschichttheorie für hohe Frequenzen.

Auf die Diskrepanz zu den Farahschen Ergebnissen wird an gegebener Stelle noch ausführlich eingegangen.

4. Grenzschichttheorie für sehr schnelle Schwingungen

Betrachten wir die für das System (3.17), (3.18) typische Dgl.

$$(i\Omega + \mathscr{E})\, P = 0 \tag{4.1}$$

im Falle hoher Frequenzen Ω.

Die einfache Vorwärtsentwicklung liefert offensichtlich $P \equiv 0$, so daß P außerhalb Wandnähe verschwindet. Somit besteht P lediglich aus zwei Grenzschichten, die zur Mitte des Strömungsfeldes hin abklingen. Um zu vermeiden, daß die beiden Grenzschichten interferieren, wird vorausgesetzt, daß die doppelte Grenzschichtdicke, die gegeben ist durch $2/\sqrt{\Omega}$, kleiner ist als die Spaltbreite:

$$\frac{2}{\sqrt{\Omega}} < R - 1\,. \tag{4.2}$$

Sei $P_l(\xi)$ die „linke" Grenzschicht oder die Grenzschicht am Rand $r = 1$. ξ hat die Bedeutung der linken inneren Variablen:

$$\xi = \sqrt{\Omega}\,(r - 1) = \frac{r-1}{\sigma}\,; \qquad r = 1 + \sigma\xi \tag{4.3}$$

mit dem kleinen Entwicklungsparameter

$$\sigma = \frac{1}{\sqrt{\Omega}} \ll 1\,. \tag{4.4}$$

Die „rechte" Grenzschicht am Rand $r = R$ wird $P_r(\eta)$ genannt, wobei η die rechte innere Variable darstellt:

$$\eta = \sqrt{\Omega}\,(R - r) = \frac{R-r}{\sigma}\,; \qquad r = R - \sigma\eta\,. \tag{4.5}$$

Das Auffinden der Grenzschichtlösungen sei nun am Beispiel der linken Grenzschicht vorgeführt. Die rechte Grenzschicht findet man mit einem völlig analogen Verfahren.

Einsetzen der inneren Variablen in (4.1) liefert:

$$\left(i + \frac{\mathrm{d}^2}{\mathrm{d}\xi^2} + \frac{\sigma}{1+\sigma\xi}\frac{\mathrm{d}}{\mathrm{d}\xi} - \frac{\sigma^2}{(1+\sigma\xi)^2}\right) P_l(\xi) = 0\,. \tag{4.6}$$

Durch Entwicklung nach σ bis zum quadratischen Glied erhält man:

$$\left(i + \frac{\mathrm{d}^2}{\mathrm{d}\xi^2} + \sigma\,[1 - \sigma\xi]\,\frac{\mathrm{d}}{\mathrm{d}\xi} - \sigma^2\right) P_l(\xi) = 0\,. \tag{4.7}$$

Mit Hilfe des Störungsansatzes

$$P_l(\xi) = P_{l,0}(\xi) + \sigma P_{l,1}(\xi) + \sigma^2 P_{l,2}(\xi) + \dots \tag{4.8}$$

entsteht das Differentialgleichungssystem:

$$iP_{l,0} + P''_{l,0} = 0\,, \qquad iP_{l,1} + P''_{l,1} = -P'_{l,0}\,, \qquad iP_{l,2} + P''_{l,2} = -P'_{l,1} + \xi P'_{l,0} + P_{l,0}\,. \tag{4.9}$$

Hier beschreibt das Symbol „'" Ableitungen nach ξ. (4.9) hat die Lösungen:

$$P_{l,0} = C_0\,\mathrm{e}^{-\sqrt{-i}\,\xi}\,, \qquad P_{l,1} = -\tfrac{1}{2}\,C_0\xi\,\mathrm{e}^{-\sqrt{-i}\,\xi} + C_1\,\mathrm{e}^{-\sqrt{-i}\,\xi}\,, \tag{4.10}$$

$$P_{l,2} = \frac{3}{8}\,C_0\xi^2\,\mathrm{e}^{-\sqrt{-i}\,\xi} - \left(\frac{1}{2}\,C_1 + \frac{3C_0}{8\sqrt{-i}}\right)\xi\,\mathrm{e}^{-\sqrt{-i}\,\xi} + C_2\,\mathrm{e}^{-\sqrt{-i}\,\xi}\,.$$

Es wurden natürlich nur die Lösungsanteile berücksichtigt, die in die Strömung hinein abklingen. Die Konstanten C_0, C_1, C_2 sind bisher noch willkürlich und erst aus den Randbedingungen des jeweils zugehörigen Randwertproblems zu bestimmen.

Auf der rechten Seite führt ein völlig analoges Verfahren zu der rechten Grenzschicht:

$$P_r(\eta) = P_{r,0} + \sigma P_{r,1} + \sigma^2 P_{r,2} + \dots\,,$$

$$P_{r,0} = D_0\,\mathrm{e}^{-\sqrt{-i}\,\eta}\,, \qquad P_{r,1} = \frac{D_0}{2R}\,\eta\,\mathrm{e}^{-\sqrt{-i}\,\eta} + D_1\,\mathrm{e}^{-\sqrt{-i}\,\eta}\,, \tag{4.11}$$

$$P_{r,2} = \frac{3}{8}\frac{D_0}{R^2}\,\eta^2\,\mathrm{e}^{-\sqrt{-i}\,\eta} - \left(\frac{3}{8\sqrt{-i}}\frac{D_0}{R^2} - \frac{D_1}{2R}\right)\eta\,\mathrm{e}^{-\sqrt{-i}\,\eta} + D_2\,\mathrm{e}^{-\sqrt{-i}\,\eta}\,.$$

4.1 Die Querströmung

Für die Querströmung F aus (3.17) gilt aufgrund der Vertauschbarkeit der Operatoren $\mathscr{E}$ und $(i\Omega + \mathscr{E})$:

$$F(r) = Ar + \frac{B}{r} + P_l(\xi) + P_r(\eta)\,, \tag{4.12}$$

mit

$$A = A_0 + \sigma A_1 + \sigma^2 A_2 + \dots\,, \qquad B = B_0 + \sigma B_1 + \sigma^2 B_2 + \dots\,. \tag{4.13}$$

Die Randbedingungen (3.17) legen sämtliche Konstanten $A_0 - D_2$ fest. Die Lösung für F wird:

$$F(r) = F_z(r) + F_l(\xi) + F_r(\eta)\,,$$

$$F_z(r) = Ar + \frac{B}{r} = \frac{r - R^2/r}{R^2 - 1} + 2\sigma R \frac{r - \frac{1}{r}(R^2 - R + 1)}{\sqrt{-i}\,(R^2 - 1)(R - 1)} +$$

$$+ i\sigma^2 \frac{(R^2 + 1)^2 r - \frac{1}{r}(R^4 - 2R^3 + 6R^2 - 2R + 1)}{(R - 1)^3 (R + 1)} + \sigma^3 F_{z,3}(r)\,, \tag{4.14}$$

$$F_l(\xi) = \frac{2R^2\sigma}{\sqrt{-i}\,(R^2 - 1)} \mathrm{e}^{-\sqrt{-i}\xi} - \frac{\sigma^2 R^2}{\sqrt{-i}\,(R^2 - 1)} \xi\, \mathrm{e}^{-\sqrt{-i}\xi} + i\sigma^2 \frac{R^3 - R^2 + 4R}{(R - 1)(R^2 - 1)} \mathrm{e}^{-\sqrt{-i}\xi}\,,$$

$$F_r(\eta) = -\frac{2\sigma}{\sqrt{-i}\,(R^2 - 1)} \mathrm{e}^{-\sqrt{-i}\eta} - \frac{\sigma^2}{\sqrt{-i}R\,(R^2 - 1)} \eta\, \mathrm{e}^{-\sqrt{-i}\eta} - i\sigma^2 \frac{1 - R + 4R^2}{R(R^2 - 1)(R - 1)} \mathrm{e}^{-\sqrt{-i}\eta}\,.$$

Der Index „z" deutet an, daß die jeweilige Funktion ein „zentraler" Lösungsanteil ist, also ein Anteil zwischen den Grenzschichten. Offenbar sind die Konstanten C_0 und D_0 zu Null geworden und die Grenzschichten beginnen erst mit dem linearen Term. Daß das so sein muß, wird sofort klar, wenn man bedenkt, daß die Ableitung von F laut den Randbedingungen (4.17) die Größenordnung 1 besitzen sollen und daß beim Nachdifferenzieren der Grenzschichtvariablen nach r ein Faktor $1/\sigma$ vorspringt.

In der Volumenstromgleichung (3.21) sind die beiden Integrale mit $\Omega = 1/\sigma^2$ multipliziert. Deshalb muß man, um die Asymptote für beliebig schnelle Schwingungen zu erhalten, bis σ^2 entwickeln. Ziel soll jedoch auch die Berechnung des in σ linearen Terms des Durchflusses sein. Deshalb muß man die Funktionen F und G bis σ^3 entwickeln. Trotzdem genügt die Kenntnis des quadratischen Effekts von F. Das hat folgenden Grund:

Im Fall, daß die Integration (3.21) über Grenzschichten auszuführen ist, gilt: $\mathrm{d}r = \sigma\,\mathrm{d}\xi$, wodurch in σ kubische Grenzschichtanteile erst den quadratischen Effekt des Volumenstromes beeinflussen können. Diese Anteile brauchen daher nicht ermittelt zu werden.

Der kubische zentrale Anteil von F, $F_{z,3}(r)$, aus (4.14), ist überflüssig, da er, wie wir im nächsten Kapitel sehen werden, bei der Realteilbildung $\mathrm{Re}\,\overline{F}G$ im Volumenstromintegral entfällt.

4.2 Die Längsströmung

Der homogene Anteil von G aus (3.18) kann dargestellt werden als:

$$G_h(r) = P_l(\xi) + P_r(\eta)\,. \tag{4.15}$$

Da es sich wieder um Grenzschichten handelt, genügt die Kenntnis von P_l und P_r bis zum quadratischen Term. Die freien Konstanten von P_l und P_r sind jetzt aus den Randvorgaben von G zu bestimmen.

Der partikuläre Lösungsanteil von G wird zerlegt in eine zentrale Lösung G_z und zwei Grenzschichten G_l und G_r, die folgende Dgln. zu erfüllen haben:

$$(i + \sigma^2\mathscr{E})\,G_z(r) = F_z(r)\left(\frac{1}{2} - \frac{R^2 - 1}{4r^2 \ln R}\right), \tag{4.16}$$

$$\left(i + \frac{\mathrm{d}^2}{\mathrm{d}\xi^2} + \sigma\,[1 - \sigma\xi]\frac{\mathrm{d}}{\mathrm{d}\xi} - \sigma^2\right)G_l(\xi) = F_l(\xi)\left(\frac{1}{2} - \frac{R^2 - 1}{4 \ln R}[1 - 2\sigma\xi + 3\sigma^2\xi^2]\right), \tag{4.17}$$

$$\left(i + \frac{\mathrm{d}^2}{\mathrm{d}\eta^2} - \frac{\sigma}{R}\left[1 + \frac{\sigma\eta}{R}\right]\frac{\mathrm{d}}{\mathrm{d}\eta} - \frac{\sigma^2}{R^2}\right)G_r(\eta) = F_r(\eta)\cdot\left(\frac{1}{2} - \frac{R^2 - 1}{4R^2 \ln R}\left[1 + 2\frac{\sigma\eta}{R} + 3\frac{\sigma^2\eta^2}{R^2}\right]\right). \tag{4.18}$$

Mit den Ansätzen

$$G_z(r) = G_{z,0} + \sigma G_{z,1} + \sigma^2 G_{z,2} + \sigma^3 G_{z,3} + \dots\,,$$

$$G_l(\xi) = G_{l,0} + \sigma G_{l,1} + \sigma^2 G_{l,2} + \dots\,, \qquad G_r(\eta) = G_{r,0} + \sigma G_{r,1} + \sigma^2 G_{r,2} + \dots \tag{4.19}$$

können die Dgln. (4.16)—(4.18) gelöst werden. Für die Grenzschichten ist mit obigem Argument wiederum die Kenntnis der quadratischen Terme ausreichend, jedoch nicht für die zentrale Lösung.

Um die Diskrepanz zwischen der vorliegenden und Farahs Arbeit deutlich zu zeigen, sei $G_z(r)$ kurz weiterdiskutiert.

Einsetzen von (4.19) in (4.16) liefert nach Abgleich in Potenzen von σ:

$$G_{z,0}(r) = -iF_{z,0}(r)\left(\frac{1}{2} - \frac{R^2-1}{4r^2 \ln R}\right), \qquad G_{z,1}(r) = -iF_{z,1}(r)\left(\frac{1}{2} - \frac{R^2-1}{4r^2 \ln R}\right),$$

$$G_{z,2}(r) = -iF_{z,2}(r)\left(\frac{1}{2} - \frac{R^2-1}{4r^2 \ln R}\right) + i\mathcal{E}G_{z,0}(r) = -iF_{z,2}(r)\left(\frac{1}{2} - \frac{R^2-1}{4r^2 \ln R}\right) + 2\,\frac{R^2}{\ln R}\cdot\frac{1}{r^5}, \tag{4.20}$$

$$G_{z,3}(r) = -iF_{z,3}(r)\left(\frac{1}{2} - \frac{R^2-1}{4r^2 \ln R}\right) + i\mathcal{E}G_{z,1}(r) = -iF_{z,3}(r)\left(\frac{1}{2} - \frac{R^2-1}{4r^2 \ln R}\right) + 4\sqrt{2}\,\frac{R(R^2-R+1)}{(R-1)\ln R}\cdot\frac{1}{r^5}.$$

Offenbar entfallen jeweils die Anteile $-iF_{z,k}(r)\left(\frac{1}{2} - \frac{R^2-1}{4r^2 \ln R}\right)$ bei der Realteilbildung $\operatorname{Re} \overline{F}G$. Jedoch der Term $i\mathcal{E}G_{z,0}(r)$ liefert einen Anteil an der Asymptoten. Da dieser Term von zweiter Ordnung in σ ist, ist, wie bereits erwähnt, die Kenntnis des kubischen Anteils von F unnötig.

Hier liegt der entscheidende Fehler der Arbeit FARAHS. Da er nach (3.23) in seiner Approximation nur gröbste Anteile berücksichtigt, hat er zumindest den quadratischen Anteil von G, $i\mathcal{E}_{z,0}G(r)$, oder wenigstens den Term seiner Theorie, der diesem entspricht, unterschlagen. Da dieser Term jedoch einen wesentlichen Beitrag zur Asymptoten liefert, ist die FARAHsche Asymptote falsch, und daher natürlich auch die gesamte Volumenstromkurve für hohe Frequenzen. Somit wird die Frage nach der Existenz des so unanschaulichen Minimums seiner Durchflußkurven völlig neu aufgeworfen und in der vorliegenden Arbeit, wie man im nächsten Kapitel sieht, dahingehend beantwortet, daß ein monotoner Übergang zwischen dem Durchfluß bei kleiner Frequenz und dem bei hoher Frequenz stattfindet.

4.3 Der Durchfluß

Bei der Bildung der Integrale (3.21) können folgende Fälle auftreten:

1. Zusammentreffen zweier zentraler Lösungen. Hier sind Integrale der Form

$$\int_1^R \operatorname{Re} \overline{F}_{z,0}(r)\, G_{z,2}(r)\, r\, dr\,, \tag{4.21}$$

auszuwerten.

2. Zusammentreffen einer zentralen Lösung mit einer Grenzschicht. Hier muß die zentrale Lösung in innere Variable umgeschrieben und entwickelt werden:

$$\int_1^R \operatorname{Re} \overline{F}_{l,1}(\xi)\, G_{z,0}(r)\, r\, dr = \int_0^\infty \operatorname{Re} \overline{F}_{l,1}(\xi)\, G_{z,0}(1+\sigma\xi)\cdot(1+\sigma\xi)\,\sigma\, d\xi\,. \tag{4.22}$$

3. Zusammentreffen zweier Grenzschichten:

$$\int_1^R \operatorname{Re} \overline{F}_{l,1}(\xi)\, G_{l,1}(\xi)\, r\, dr = \int_0^\infty \operatorname{Re} \overline{F}_{l,1}(\xi)\, G_{l,1}(\xi)\cdot(1+\sigma\xi)\,\sigma\, d\xi\,. \tag{4.23}$$

Da nach Voraussetzung (4.2) linke und rechte Grenzschicht nicht interferieren, entsteht bei deren Aufeinandertreffen natürlich eine Null:

$$\int_1^R \operatorname{Re} \overline{F}_{l,1}(\xi)\, G_{r,1}(\eta)\, r\, dr = 0\,. \tag{4.24}$$

Die Ausführung der beschriebenen Verfahren liefert folgende explizite Darstellung des Volumenstromes:

$$\frac{\dot V - \dot V_0}{\dot V_0 \varepsilon^2} = \frac{1}{R^4 - 1 - \frac{(R^2-1)^2}{\ln R}}\left\{-\frac{R^2-1}{6R^2 \ln^2 R}(R^4+R^2+1) + \frac{R^2+1}{\ln R} - \frac{2R^2}{R^2-1} + \right.$$
$$+\frac{1}{\sqrt{\Omega}}\left[-\frac{R+1}{12R^3 \ln^2 R}(-7R^6+7R^5+R^4+7R^3+R^2+7R-7) - \right.$$
$$\left.\left. - \frac{R^4-R^3-3R^2-R+1}{R(R-1)\ln R} - 3\,\frac{R(R^2-R+1)}{(R^2-1)(R-1)}\right]\right\}. \tag{4.25}$$

Fig. 2 zeigt diese Kurven für schnelle Schwingungen, während die Kurven kleiner Frequenz FARAHS Arbeit entnommen sind.

Offenbar gehen die Volumenstromkurven monoton über von den Parabeln sehr langsamer Schwingungen in die Hyperbeln sehr schneller Schwingungen. Bei langsamen Schwingungen tritt Durchflußerhöhung, bei schnellen Schwingungen Durchflußminderung gegenüber dem Fall des ruhenden Innenzylinders ein.

Diese Effekte sind um so stärker, je enger der Spalt ist, was man leicht verstehen kann: Für große Spaltbreiten klingen die vom Innenzylinder verursachten Störungen rasch nach außen hin ab, so daß im Hauptteil des Spaltes in etwa der gleiche Volumenstrom fließt, wie im Fall des ruhenden Innenzylinders. Die Durchflußänderungen bleiben im wesentlichen auf das Gebiet in der Nähe des Innenzylinders beschränkt.

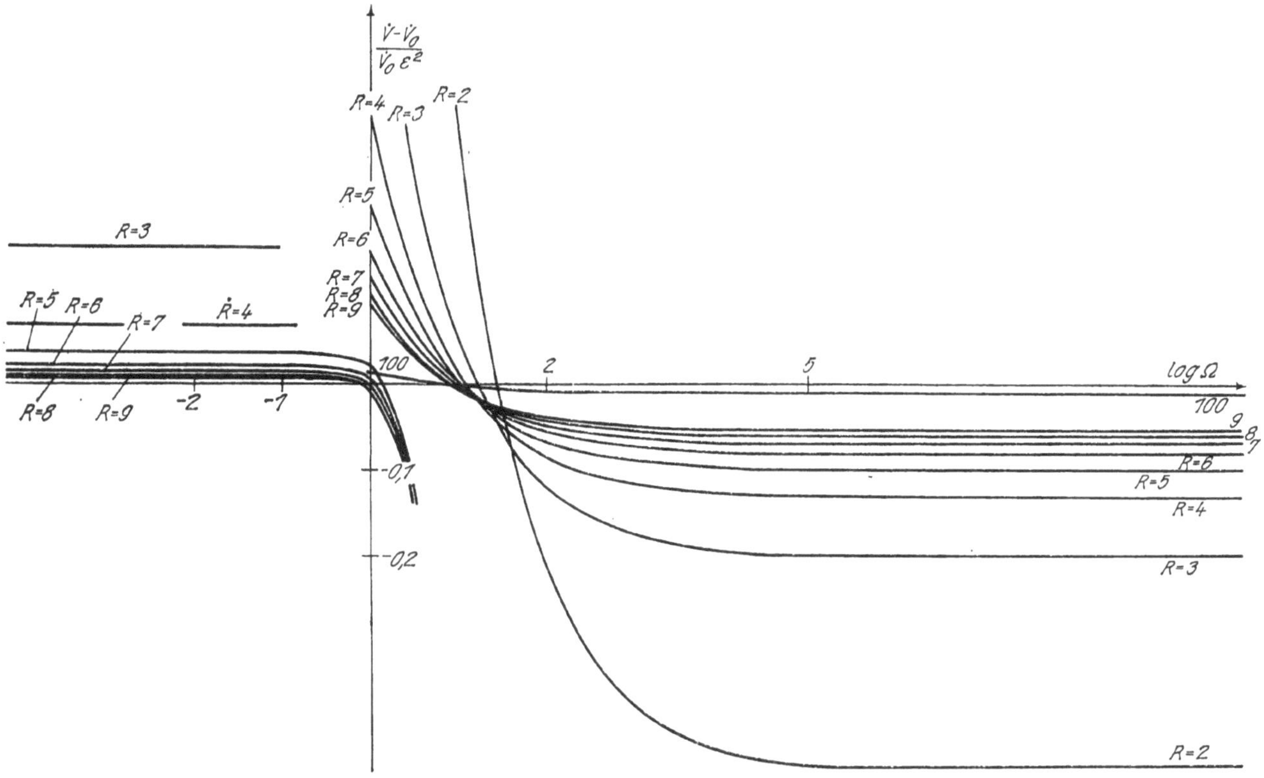

Fig. 2

Literatur

1 ABRAMOWITZ, M.; STEGUN, I. A., Handbook of Mathematical Functions with Formulas, Graphs and Mathematical Tables, Dover Public. Inc., New York.
2 BATCHELOR, G. K., Fluid Dynamics, Chap. 5.13: Oscillating boundary layers, Cambridge 1967.
3 VAN DYKE, M., Perturbation Methods in Fluid Mechanics, Academic Press, New York and London, 1964.
4 FARAH, M. N., Laminare instationäre Strömung im Ringraum zwischen zwei Rohren, von denen das innere harmonisch schwingt, ZAMM 61, 365–377 (1981).
5 SCHLICHTING, H., Grenzschichttheorie, 5. Aufl. Kap. XV, e: Periodische Grenzschichtströmungen; e 1: Der oszillierende Zylinder, Karlsruhe 1965.
6 TIEDT, W., Berechnung des laminaren und turbulenten Reibungswiderstandes konzentrischer und exzentrischer Ringspalte, Chemiker-Zeitung/Chemische Apparatur 90, 1–32 (1966) und 91, 309–312 (1967).
7 WAMBSGANSS, M. W.; CHEN, S. S.; JENDRZEYCZYK, J. A., Added Mass and Damping of a Vibrating Rod in Confined Viscous Fluids, Argonne National Laboratory Rep. ANL-CT-75-08 (1974).

Eingereicht am 26. 11. 1981

Anschrift: Dipl.-Ing. HARALD MÜLLER, Technische Hochschule, Institut für Mechanik, Hochschulstraße 1, D 6100 Darmstadt, BRD

ZAMM 62, 511 – 520 (1982)

HELMUT F. BAUER

Flow in a Cylindrical Tube with sudden Change of Suction Across the Wall

Es wird die Übergangsströmung in einem zylindrischen Rohr mit Kreisquerschnitt und konstanter Absaugung über die permeable Wand bezüglich einer plötzlichen Änderung der Absaugung an der Wand bestimmt. Dabei werden Geschwindigkeits-, Spannungs- und Druckverteilung analytisch bestimmt. Aus den Ergebnissen können einige spezielle Fälle, die für technische wie auch biologische Prozesse (Resorption) von Bedeutung sind, gewonnen werden.

In a permeable circular cylindrical tube with constant suction across the wall the constant radial flow velocity at the wall is suddenly changed to another constant value. The transient flow behavior, such as the velocity-, stress- and pressure distribution are presented analytically. Many special cases of interest to engineering and biological processes may be obtained.

Определяется переходное течение в цилиндрической трубе круглого сечения с постоянным отсасыванием сквозь проницаемую стенку относительно внезапной перемены отсасывания вдоль стенке. При этом в аналитическом виде определяются распределения скоростей, напряжений и давления. Можно получить из результатов несколько частных случаев являющих важным и для технических и для биологических процессов.

1. Introduction

Injection or suction across the wall of a liquid-carrying tube have many applications in engineering. Injection may be used for cooling purposes in engines or power plants, while suction or reabsorption across the wall may have many applications in bio-engineering. In many biological fluid flow systems liquid is increased or decreased by reabsorption processes across the wall of the tube for reasons of exchange processes [1]. This may be the case in blood vessels of the circulation system or in the nephron, where the largest amount of filtered volume through the glomerulum is reabsorbed in the tubulus. For the determination of the mass transport in the mentioned systems as well as those of the artificial kidney the convective part of the mass transport plays an immediate important role. In most practical cases the reabsorption processes across the wall decrease or increase the liquid flow in the tube in a linear fashion. This is expressed by a constant flow across the wall per unit length. A sudden change of this constant suction may be quite important for the behavior of the concentration of a material component dissolved in the liquid. The purpose of this paper is therefore the investigation of the transient flow behavior of such a system with permeable wall due to the sudden change of the suction velocity at the wall of the tube. The case of steady linear [2, 3] and nonlinear flow of Newtonian liquid in a circular tube with constant and exponentially decaying reabsorption, as well as the linearized pulsatile case have been treated before [4]. The case of sudden suction change, however, has not been treated. This transient case shall be investigated here, where the velocity-, pressure- and stress-distribution shall be obtained for all time regimes. Various special cases of interest may be obtained from the analytical results. It may be noticed that a sudden increase of suction decreases the flow in axial direction suddenly, where the flattened axial velocity profil assumes with increasing time the parabolic form. In the case of a sudden decrease of suction at the wall of the tube a backflow occurs for large decreases, a case that may be of some interest in biological systems.

2. Formulation of the Problem

We consider a long circular cylindrical tube of radius a in which a homogeneous incompressible Newtonian viscous liquid flows in axial direction (z direction) and through which wall reabsorption occurs in a constant fashion. This means, that at the location $z = 0$ the flow volume per unit time is given by $\dot{V}_0 =$ constant, while at the location z it shall be $\dot{V}(z)$, i.e., $\dot{V}(z) = \dot{V}_0 - 2\pi a u_0 z$, where u_0 is the radial velocity component at the wall of the tube. This yields for the outflow of volume per time unit through the wall of the tube the value $\dot{V}_{\text{out}} = \dot{V}_0 - \dot{V}(z) = 2\pi a u_0 z$ (Figure 1). This flow case exists for all time values $t < 0$. At $t = 0$ a sudden change in the radial velocity of the wall occurs to a value of magnitude U_0, which means that the outflow suddenly changes to $2\pi a U_0 z$. The problem to be solved is the transient flow behavior in the tube for $t \geqq 0$ (Figure 2).

The equations governing the axisymmetric flow of a viscous liquid are the Navier-Stokes equations, which are in linearized form given as

$$\frac{\partial u}{\partial t} + \frac{1}{\varrho}\frac{\partial p}{\partial r} = \nu\left[\frac{\partial^2 u}{\partial r^2} + \frac{1}{r}\frac{\partial u}{\partial r} - \frac{u}{r^2} + \frac{\partial^2 u}{\partial z^2}\right] \tag{1}$$

in radial direction, and

$$\frac{\partial w}{\partial t} + \frac{1}{\varrho}\frac{\partial p}{\partial z} = \nu\left[\frac{\partial^2 w}{\partial r^2} + \frac{1}{r}\frac{\partial w}{\partial r} + \frac{\partial^2 w}{\partial z^2}\right] \tag{2}$$

in axial direction, and the equation of continuity

$$\frac{\partial u}{\partial r} + \frac{u}{r} + \frac{\partial w}{\partial z} = 0\,. \tag{3}$$

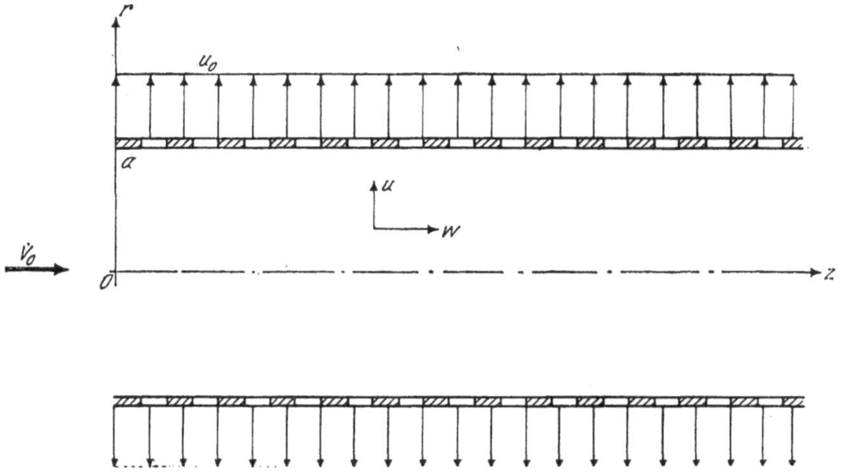

Fig. 1. Geometry and coordinates of system

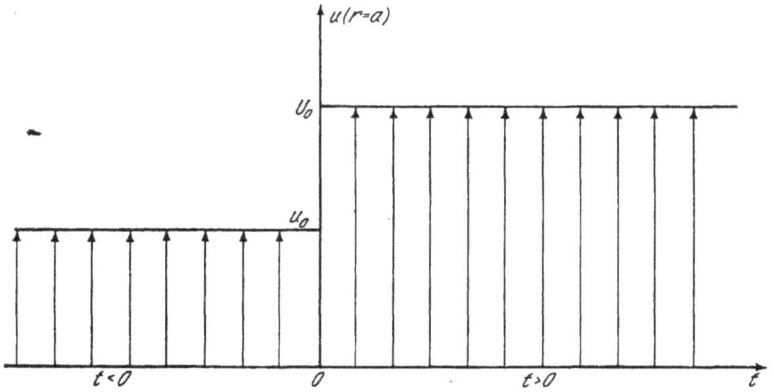

Fig. 2. Sudden change of wall suction velocity at $t = 0$

In these equations, r and z are cylindrical polar coordinates, with r measured in radial direction from the centerline of the tube and z measured in axial direction along this line. The velocity components in the direction of increasing r and z, respectively, are denoted by u and w. The value ϱ denotes the constant mass density of the liquid, $\nu = \eta/\varrho$ the kinematic viscosity, and p the pressure. The problem is to find for constant reabsorption the velocity distribution and the pressure for all times, i.e. for $t < 0$ and $t \geqq 0$. The velocity component u is due to the reabsorption process through the wall and is vanishing on the centerline of the tube. Also the axial velocity component w exhibits a finite value on the centerline. If there is no reabsorption, u is vanishing throughout and a simple Poiseuille flow would be obtained. The boundary conditions for the transient problem are given by

$$w(a, t) = 0 \quad \text{at the wall} \quad r = a \tag{4}$$

and

$$u(a, t) = \left\{ \begin{matrix} u_0 & \text{for} & t < 0 \\ U_0 & \text{for} & t \geqq 0 \end{matrix} \right\} \text{ at the wall } r = a\,. \tag{5}$$

By introducing the stream function $\psi(r, z, t)$ such that the equation of continuity (3) is satisfied identically, i.e., in terms of which

$$u = \frac{1}{r}\frac{\partial \psi}{\partial z}, \qquad w = -\frac{1}{r}\frac{\partial \psi}{\partial r}$$

and eliminating p from the Navier-Stokes equations by differentiating equation (1) with respect to z and equation (2) with respect to r and subtracting them, yields the partial differential equation for the stream function:

$$\overline{\Delta}\left[\overline{\Delta}\psi - \frac{1}{\nu}\frac{\partial \psi}{\partial t}\right] = 0 \tag{7}$$

where

$$\overline{\Delta} \equiv \frac{\partial^2}{\partial r^2} - \frac{1}{r}\frac{\partial}{\partial r} + \frac{\partial^2}{\partial z^2}.$$

3. Method of Solution

The problem shall be solved for two time regions, i.e., the time regions $t < 0$ and $t \geqq 0$.

3.1. Flow during $t < 0$

In this case the result for the flow may be obtained by solving [1]

$$\bar{\Delta}\bar{\Delta}\psi_0^{(0)} = 0 \tag{8}$$

with the boundary conditions

$$u = u_0 \quad \text{and} \quad w = 0 \quad \text{at the wall} \quad r = a\,. \tag{9}$$

Observing also the fact that u vanishes on the axis $r = 0$ and w exhibits a finite value there, with $\dot{V} = \dot{V}_0(1 - \lambda_0 z)$, $\lambda_0 \equiv 2\pi a u_0/\dot{V}_0$ and $\psi_0^{(0)} = \varphi_0(r)\,(1 - \lambda_0 z)$ the stream function is given by the expression

$$\psi_0^{(0)}(r, z) = [A_0^{(0)} r^2 + B_0^{(0)} r^4]\,(1 - \lambda_0 z)$$

which with the boundary conditions (9) yields

$$\psi_0^{(0)}(r, z) = -\frac{u_0 a}{\lambda_0}\left(\frac{r}{a}\right)^2\left[2 - \left(\frac{r}{a}\right)^2\right](1 - \lambda_0 z) \quad \text{for} \quad t < 0\,. \tag{10}$$

The velocity distribution is for $t < 0$

$$u^{(0)}(r, z) = u_0\frac{r}{a}\left[2 - \left(\frac{r}{a}\right)^2\right], \qquad w^{(0)}(r, z) = \frac{2\dot{V}_0}{\pi a^2}\left[1 - \left(\frac{r}{a}\right)^2\right](1 - \lambda_0 z)\,. \tag{11}$$

The pressure distribution is obtained by integrating the stationary Navier-Stokes equation (1) and (2), i.e.,

$$p^{(0)} = C_1(z) + \eta\int\left[u_{rr}^{(0)} + \frac{u_r^{(0)}}{r} - \frac{u^{(0)}}{r^2} + u_{zz}^{(0)}\right]\mathrm{d}r\,, \quad p^{(0)} = C_2(r) + \eta\int\left[w_{rr}^{(0)} + \frac{1}{r}w_r^{(0)} + w_{zz}^{(0)}\right]\mathrm{d}z$$

and results in

$$p^{(0)}(r, z) = C_1(z) - \frac{4u_0\eta}{a}\left(\frac{r}{a}\right)^2, \qquad p^{(0)}(r, z) = C_2(r) - \frac{8u_0\eta}{\lambda_0 a^3}z\left(1 - \frac{\lambda_0}{2}z\right).$$

Thus the pressure distribution for $t < 0$ is given by

$$p^{(0)}(r, z) = p(0, 0) - \frac{4u_0\eta}{a}\left\{\left(\frac{r}{a}\right)^2 + \frac{4z}{\lambda_0 a^2}\left(1 - \frac{\lambda_0}{2}z\right)\right\}. \tag{12}$$

3.2. Flow in the time region $t \geqq 0$

In this time region a transient occurs, which leads the above flow into a similar flow with U_0 instead of u_0. Dividing the stream function into two parts, one of which represents the steady (final) flow case (ψ_0), while the other describes the transient behavior of the flow, it is:

$$\psi(r, z, t) = \psi_0(r, z) + \psi_1(r, z, t)\,. \tag{13}$$

Introducing this into equation (7) yields

$$\bar{\Delta}\bar{\Delta}\psi_0 + \bar{\Delta}\left[\bar{\Delta}\psi_1 - \frac{1}{\nu}\frac{\partial\psi_1}{\partial t}\right] = 0$$

where the stationary part exhibits the solution

$$\psi_0(r, z) = (A_0 r^2 + B_0 r^4)\,(1 - \bar{\lambda}_0 z) \tag{14}$$

with $\bar{\lambda}_0 = 2\pi a U_0/\dot{V}_0$, and the instationary part yields

$$\psi_1(r, z, t) = \left\{C r J_1(\mu r) + \frac{A}{\mu^2}r^2\right\}(1 - \bar{\lambda}_0 z)\,\mathrm{e}^{-\mu^2\nu t}\,. \tag{15}$$

The complete solution for $t > 0$ is then given by

$$\psi(r, z, t) = [A_0 r^2 + B_0 r^4]\,(1 - \bar{\lambda}_0 z) + \left\{C r J_1(\mu r) + \frac{A}{\mu^2}r^2\right\}(1 - \bar{\lambda}_0 z)\,\mathrm{e}^{-\mu^2\nu t}\,. \tag{16}$$

Satisfying the boundary conditions (4) and (5) results in

$$A_0 + 2a^2 B_0 = 0\,, \qquad C[J_1(\mu a) + \mu a J_1'(\mu a)] + 2a\frac{A}{\mu^2} = 0$$

and

$$A_0 + a^2 B_0 = -\frac{U_0}{\bar{\lambda}_0 a}\,, \qquad C J_1(\mu a) + \frac{A}{\mu^2} = 0\,.$$

Therefore $A_0 = -2U_0/\bar{\lambda}_0 a$; $B_0 = U_0/\bar{\lambda}_0 a^3$; $A/\mu^2 = -CJ_1(\mu a)/a$ while the condition $\mu a J_1'(\mu a) - J_1(\mu a) = 0$. This is equivalent with $\mu a J_2(\mu a) = 0$ and is satisfied for $\mu_m a = \varepsilon_m$, where ε_m are the roots of the Bessel function of first kind and second order, i.e., $J_2(\varepsilon_m) = 0$. They are [5]

$$\varepsilon_1 = 5.1356\,, \quad \varepsilon_2 = 8.4172\,, \quad \varepsilon_3 = 11.6198\,, \quad \varepsilon_4 = 14.7960\,, \quad \varepsilon_5 = 17.9598\,, \ldots$$

The stream function for $t \geqq 0$ is then

$$\psi(r, z, t) = -(1 - \bar{\lambda}_0 z)\frac{U_0 a}{\bar{\lambda}_0}\left(\frac{r}{a}\right)^2\left[2 - \left(\frac{r}{a}\right)^2\right] + \sum_{m=1}^{\infty} aC_m(1 - \bar{\lambda}_0 z)\left[\frac{r}{a} J_1\left(\varepsilon_m \frac{r}{a}\right) - \left(\frac{r}{a}\right)^2 J_1(\varepsilon_m)\right] e^{-(\varepsilon_m^2/a^2)\nu t}\,. \tag{17}$$

For the determination of the constants C_m, one has to observe the condition at $t = 0$, where $u_0^{(0)}(r, z) = u(r, z, 0)$. This yields with equation (6), (11) and (17) the expression

$$(U_0 - u_0)\left[2\left(\frac{r}{a}\right) - \left(\frac{r}{a}\right)^3\right] = \sum_{n=1}^{\infty} C_n \bar{\lambda}_0 \left\{J_1\left(\varepsilon_n \frac{r}{a}\right) - \frac{r}{a} J_1(\varepsilon_n)\right\}. \tag{18}$$

Expanding the functions $\frac{r}{a}$, $\left(\frac{r}{a}\right)^3$ and $J_1\left(\varepsilon_n \frac{r}{a}\right)$ into Bessel-Fourier series with respect to $J_2\left(\varepsilon_m \frac{r}{a}\right)$ yields with [6]:

$$\int \xi^k J_\nu(\xi)\, d\xi = \xi^k \frac{\Gamma\left(\frac{k+\nu+1}{2}\right)}{\Gamma\left(\frac{\nu-k+1}{2}\right)} \sum_{\mu=0}^{\infty} \frac{(2\mu+\nu+1)\,\Gamma\left(\mu + \frac{\nu-k+1}{2}\right)}{\Gamma\left(\mu + \frac{\nu+k+3}{2}\right)} J_{2\mu+\nu+1}(\xi) \tag{19}$$

for

$$\frac{r}{a} = \sum_{m=1}^{\infty} \alpha_m J_2\left(\varepsilon_m \frac{r}{a}\right) \quad \text{with} \quad \alpha_m = \frac{2}{J_3^2(\varepsilon_m)} \int_0^1 \xi^2 J_2(\varepsilon_m \xi)\, d\xi\,, \tag{20}$$

$$\left(\frac{r}{a}\right)^3 = \sum_{m=1}^{\infty} \beta_m J_2\left(\varepsilon_m \frac{r}{a}\right) \quad \text{with} \quad \beta_m = \frac{2}{J_3^2(\varepsilon_m)} \int_0^1 \xi^4 J_2(\varepsilon_m \xi)\, d\xi\,, \tag{21}$$

$$J_1\left(\varepsilon_n \frac{r}{a}\right) = \sum_{m=1}^{\infty} \gamma_{mn} J_2\left(\varepsilon_m \frac{r}{a}\right) \quad \text{with} \quad \gamma_{mn} = \frac{2}{J_3^2(\varepsilon_m)} \int_0^1 \xi J_1(\varepsilon_n \xi)\, J_2(\varepsilon_m \xi)\, d\xi\,. \tag{22}$$

With equation (19) the values α_m and β_m are obtained to be

$$\alpha_m = \frac{12}{\varepsilon_m J_3^2(\varepsilon_m)} \sum_{\mu=0}^{\infty} \frac{J_{2\mu+3}(\varepsilon_m)}{(2\mu+1)(2\mu+5)}\,, \qquad \beta_m = -\frac{60}{\varepsilon_m J_3^2(\varepsilon_m)} \sum_{\mu=0}^{\infty} \frac{J_{2\mu+3}(\varepsilon_m)}{(2\mu+7)(2\mu+5)(4\mu^2-1)}\,. \tag{23}$$

Expanding $J_1\left(\varepsilon_n \frac{r}{a}\right)$ into its power series and integrating term by term, using equation (19), results finally in the expression

$$\gamma_{mn} = \frac{2}{\varepsilon_m J_3^2(\varepsilon_m)} \sum_{\nu=0}^{\infty} \frac{(-1)^\nu \varepsilon_n^{2\nu+1}}{2^{2\nu}\nu!\,(\nu+1)\,\nu} \left\{\frac{3}{2\nu+5} J_3(\varepsilon_m) + \right.$$
$$\left. + \sum_{\mu=1}^{\infty} \frac{(-1)^\mu (2\mu+3)(2\nu-2\mu+1)(2\nu-2\mu+3)\ldots(2\nu-1)}{(2\mu+2\nu+5)(2\mu+2\nu+3)\ldots(2\nu+5)} J_{2\mu+3}(\varepsilon_m)\right\}. \tag{24}$$

Introducing these results into equation (18) and comparing coefficients of $J_2\left(\varepsilon_m \frac{r}{a}\right)$ yields an infinite set of algebraic equations for the determination of the infinite number of unknowns C_m. They are (with $\bar{\lambda}_0 C_n \equiv C_n^*$)

$$\sum_{n=1}^{\infty} C_n^*[\gamma_{mn} - \alpha_m J_1(\varepsilon_n)] = (U_0 - u_0)\,[2\alpha_m - \beta_m] \tag{25}$$

for $m = 1, 2, 3, \ldots$ This system may be solved approximately for a finite number of unknowns $C_1^*, C_2^*, \ldots, C_N^*$. The stream function is then given by the expression

$$\psi(r, z, t) = -(1 - \bar{\lambda}_0 z)\frac{U_0}{\bar{\lambda}_0} a \left(\frac{r}{a}\right)^2\left[2 - \left(\frac{r}{a}\right)^2\right] +$$
$$+ \sum_{m=1}^{\infty} \frac{aC_m^*}{\bar{\lambda}_0}(1 - \bar{\lambda}_0 z)\left[\left(\frac{r}{a}\right) J_1\left(\varepsilon_m \frac{r}{a}\right) - \left(\frac{r}{a}\right)^2 J_1(\varepsilon_m)\right] e^{-(\varepsilon_m^2/a^2)\nu t} \quad \text{for} \quad t \geqq 0\,. \tag{26}$$

The velocity distribution is given by

$$u(r, z, t) = \frac{1}{r}\frac{\partial\psi}{\partial z} = U_0 \frac{r}{a}\left[2 - \left(\frac{r}{a}\right)^2\right] - \sum_{m=1}^{\infty} C_m^* \left[J_1\left(\varepsilon_m \frac{r}{a}\right) - \frac{r}{a} J_1(\varepsilon_m)\right] e^{-(\varepsilon_m^2/a^2)\nu t} \quad \text{for} \quad t \geqq 0 \tag{27}$$

and

$$w(r, z, t) = -\frac{1}{r}\frac{\partial\psi}{\partial r} = \frac{2\dot{V}_0}{\pi a^2}\left[1 - \frac{r^2}{a^2}\right](1 - \bar{\lambda}_0 z) - \sum_{m=1}^{\infty} \frac{C_m^*}{a\bar{\lambda}_0}(1 - \bar{\lambda}_0 z) \times$$

$$\times \left[\varepsilon_m J_0\left(\varepsilon_m \frac{r}{a}\right) - 2J_1(\varepsilon_m)\right] e^{-(\varepsilon_m^2/a^2)\nu t} \quad \text{for} \quad t \geqq 0 . \tag{28}$$

The pressure distribution is obtained from

$$p(r, z, t) = C_1(z) + \eta \int \left\{u_{rr} + \frac{u_r}{r} - \frac{u}{r^2} + u_{zz} - \frac{1}{\nu} u_t\right\} dr$$

and

$$p(r, z, t) = C_2(r) + \eta \int \left\{w_{rr} + \frac{w_r}{r} + w_{zz} - \frac{1}{\nu} w_t\right\} dz$$

and yields

$$p(r, z, t) = p(0, 0, t) - \frac{4U_0\eta}{a}\left(\frac{r}{a}\right)^2 - \frac{8\eta\dot{V}_0}{\pi a}\left(z - \frac{\bar{\lambda}_0}{2} z^2\right) + \left\{\left(\frac{r}{a}\right)^2 + \frac{4(1 - \bar{\lambda}_0 z)}{a^2\bar{\lambda}_0}\right\} \sum_{m=1}^{\infty} \frac{\eta\varepsilon_m^2 C_m^*}{2a} J_1(\varepsilon_m) e^{-(\varepsilon_m^2/a^2)\nu t}. \tag{29}$$

The mean pressure over the cross section is obtained from

$$p_{\text{mean}}(z, t) = \frac{2}{a^2}\int_0^a r p(r, z, t)\, dr$$

and yields

$$p_{\text{mean}}^{(0)}(z) = p(0, 0) - \frac{4u_0\eta}{a}\left\{\frac{1}{2} + \frac{4z}{\lambda_0 a^2}\left(1 - \frac{\lambda_0 z}{2}\right)\right\} \quad \text{for} \quad t < 0 \tag{30}$$

and

$$p_{\text{mean}}(z, t) = p(0, 0, t) - \frac{2U_0\eta}{a} - \frac{8\eta\dot{V}_0 z}{\pi a^4}\left(1 - \frac{\bar{\lambda}_0 z}{2}\right) + \left\{\frac{1}{2} + \frac{4z}{a^2\bar{\lambda}_0}\left(1 - \frac{\bar{\lambda}_0 z}{2}\right)\right\} \sum_{m=1}^{\infty} \frac{C_m^*\eta\varepsilon_m^2}{2a} J_1(\varepsilon_m) e^{-(\varepsilon_m^2/a^2)\nu t}$$

$$\text{for} \quad t \geqq 0. \tag{31}$$

At $z = 0$ this is

$$p_{\text{mean}}^{(0)}(0) = p(0, 0) - \frac{2u_0\eta}{a} \quad \text{for} \quad t < 0$$

and

$$p_{\text{mean}}(0, t) = p(0, 0, t) - \frac{2U_0\eta}{a} + \frac{\eta}{4a}\sum_{m=1}^{\infty} C_m^*\varepsilon_m^2 J_1(\varepsilon_m) e^{-(\varepsilon_m^2/a^2)\nu t} \quad \text{for} \quad t \geqq 0 .$$

From the above results it may be seen that the flow yields for $U_0 = u_0 (C_m^* = 0)$, i.e., with no reabsorption change at the time $t = 0$, the expression for $t < 0$, which is then valid also for $t \geqq 0$. As special cases one may also consider those where the reabsorption stops at $t = 0$, which means $U_0 = 0$, or the one, where the reabsorption was zero ($u_0 = 0$) for $t < 0$.

In the first case the velocity distribution yields:

$$u(r, z) = u_0 \frac{r}{a}\left[2 - \left(\frac{r}{a}\right)^2\right] \quad \text{for} \quad t < 0 ,$$

$$u(r, z, t) = -\sum_{m=1}^{\infty} C_m^* \left[J_1\left(\varepsilon_m \frac{r}{a}\right) - \frac{r}{a} J_1(\varepsilon_m)\right] e^{-(\varepsilon_m^2/a^2)\nu t} \quad \text{for} \quad t \geqq 0 ,$$

and

$$w(r, z) = \frac{2\dot{V}_0}{\pi a^2}\left[1 - \left(\frac{r}{a}\right)^2\right](1 - \lambda_0 z) \quad \text{for} \quad t < 0 ,$$

$$w(r, z, t) = \frac{2\dot{V}_0}{\pi a^2}\left[1 - \left(\frac{r}{a}\right)^2\right] - \sum_{m=1}^{\infty} \frac{C_m^*}{a}\left[\varepsilon_m J_0\left(\varepsilon_m \frac{r}{a}\right) - 2J_1(\varepsilon_m)\right] e^{-(\varepsilon_m^2/a^2)\nu t} \quad \text{for} \quad t \geqq 0 .$$

As time elapses the axial velocity approaches that of Poiseuille flow.

In the case that initially ($t < 0$) there was no reabsorption ($u_0 = 0$), we obtain:

$$u(r, z) = 0 \quad \text{for} \quad t < 0\,,$$

$$u(r, z, t) = U_0 \frac{r}{a}\left[2 - \left(\frac{r}{a}\right)^2\right] - \sum_{m=1}^{\infty} C_m^* \left[J_1\left(\varepsilon_m \frac{r}{a}\right) - \frac{r}{a} J_1(\varepsilon_m)\right] e^{-(\varepsilon_m^2/a^2)\nu t} \quad \text{for} \quad t \geqq 0$$

and

$$w(r, z) = \frac{2\dot{V}_0}{\pi a^2}\left[1 - \left(\frac{r}{a}\right)^2\right] \quad \text{for} \quad t < 0\,,$$

$$w(r, z, t) = \frac{2\dot{V}_0}{\pi a^2}\left[1 - \left(\frac{r}{a}\right)^2\right](1 - \bar{\lambda}_0 z) - \frac{1}{a}\sum_{m=1}^{\infty} C_m (1 - \bar{\lambda}_0 z)\left[\varepsilon_m J_0\left(\varepsilon_m \frac{r}{a}\right) - 2 J_1(\varepsilon_m)\right] e^{-(\varepsilon_m^2/a^2)\nu t}$$

for $t \geqq 0$.

It may be seen from the previous results that for increasing time $t \to \infty$, the final steady-state condition is obtained.

4. Normal- and shear stresses

The normal and shear stresses are obtained from the expressions

$$\sigma_r = -2\eta \frac{\partial u}{\partial r} \qquad \tau_{rz} = -\eta\left[\frac{\partial u}{\partial z} + \frac{\partial w}{\partial r}\right]$$

$$\sigma_\theta = -2\eta \frac{u}{r} \qquad \tau_{r\theta} = \tau_{\theta z} = 0$$

$$\sigma_z = -2\eta \frac{\partial w}{\partial z}$$

and yield with equations (11), (12) and (27), (28) the following results:

$$\sigma_r = \begin{cases} -\dfrac{2\eta u_0}{a}\left[2 - 3\left(\dfrac{r}{a}\right)^2\right] & \text{for } t < 0 \\ -\dfrac{2\eta U_0}{a}\left[2 - 3\left(\dfrac{r}{a}\right)^2\right] + \dfrac{2\eta}{a}\sum\limits_{m=1}^{\infty} C_m^*\left[\varepsilon_m J_1'\left(\varepsilon_m \dfrac{r}{a}\right) - J_1(\varepsilon_m)\right] e^{-(\varepsilon_m^2/a^2)\nu t} & \text{for } t \geqq 0 \end{cases} \tag{32}$$

$$\sigma_\theta = \begin{cases} -\dfrac{2\eta u_0}{a}\left[2 - \left(\dfrac{r}{a}\right)^2\right] & \text{for } t < 0 \\ -\dfrac{2\eta U_0}{a}\left[2 - \left(\dfrac{r}{a}\right)^2\right] + \dfrac{2\eta}{a}\sum\limits_{m=1}^{\infty} C_m^*\left[\dfrac{J_1\left(\varepsilon_m \dfrac{r}{a}\right)}{(r/a)} - J_1(\varepsilon_m)\right] e^{-(\varepsilon_m^2/a^2)\nu t} & \text{for } t \geqq 0 \end{cases} \tag{33}$$

$$\sigma_z = \begin{cases} \dfrac{8\eta u_0}{a}\left[1 - \left(\dfrac{r}{a}\right)^2\right] & \text{for } t < 0\,, \\ \dfrac{8\eta U_0}{a}\left[1 - \left(\dfrac{r}{a}\right)^2\right] + \dfrac{2\eta}{a}\sum\limits_{m=1}^{\infty} C_m^*\left[\varepsilon_m J_0\left(\varepsilon_m \dfrac{r}{a}\right) - 2J_1(\varepsilon_m)\right] e^{-(\varepsilon_m^2/a^2)\nu t} & \text{for } t \geqq 0 \end{cases} \tag{34}$$

and

$$\tau_{rz} = \begin{cases} \dfrac{4\eta \dot{V}_0}{\pi a^3}\left(\dfrac{r}{a}\right)[1 - \lambda_0 z] & \text{for } t < 0\,, \\ \dfrac{4\eta \dot{V}_0}{\pi a^3}\left(\dfrac{r}{a}\right)[1 - \bar{\lambda}_0 z] - \dfrac{\eta}{a^2}\sum\limits_{m=1}^{\infty} \dfrac{C_m^*}{\bar{\lambda}_0}[1 - \bar{\lambda}_0 z]\, \varepsilon_m^2\, J_1\left(\varepsilon_m \dfrac{r}{a}\right) e^{-(\varepsilon_m^2/a^2)\nu t} & \text{for } t \geqq 0\,. \end{cases} \tag{35}$$

5. Numerical Evaluation

In order to determine the transient velocity distribution due to any suction change at the wall of the tube from the constant value u_0 to a value U_0 of constant magnitude, the constants $C_m^*/(U_0 - u_0)$ had to be determined numerically from the infinite system of linear equations (25). This system was truncated to M equations and solved numerically. The results are presented for $M = 20$ in the Table. It may be seen that the coefficients $C_m^*/(U_0 - u_0)$ alter in sign. In order to obtain some information about the magnitude of these coefficients the amount of the equations for their determination has been increased from $m = 5$ to 20. Only the first five coefficients are presented in the Table. It may be noticed that the convergence is very slow. It is, however, believed that the accuracy is sufficiently good for

Table: Numerical Value of Constants C_m^* (see equ. (25))

M	$C_1^*/(U_0 - u_0)$	$C_2^*/(U_0 - u_0)$	$C_3^*/(U_0 - u_0)$	$C_4^*/(U_0 - u_0)$	$C_5^*/(U_0 - u_0)$	m	$C_m^*/(U_0 - u_0)$
5	0.80078193	−0.50947561	0.91343736	−0.35783818	2.32108922	1	0.53311153
6	0.68402410	−0.63489082	0.41009188	−0.93239850	0.30450026	2	−0.32439488
7	0.69977390	−0.45431284	0.60942109	−0.35130551	0.95484303	3	0.24901558
8	0.63738239	−0.51467695	0.37059515	−0.61430162	0.31031657	4	−0.25270715
9	0.64415636	−0.41417928	0.47581340	−0.32376356	0.62784349	5	0.20903266
10	0.60575009	−0.44797282	0.33768592	−0.47108149	0.29176446	6	−0.26620889
11	0.60882734	−0.38466322	0.40082493	−0.29737322	0.47759277	7	0.19797906
12	0.58298911	−0.40547469	0.31163521	−0.38970444	0.27090676	8	−0.30710515
13	0.58434231	−0.36228384	0.35279714	−0.27496245	0.39120664	9	0.19211602
14	0.56586198	−0.37597363	0.29091868	−0.33724407	0.25190091	10	−0.36804969
15	0.56634591	−0.34481371	0.31936856	−0.25632494	0.33512826	11	0.18540778
16	0.55252227	−0.35426473	0.27419135	−0.30059757	0.23539637	12	−0.45244232
17	0.55254659	−0.33083111	0.29473167	−0.24079286	0.29578882	13	0.17541080
18	0.54184597	−0.33760100	0.26046069	−0.27353151	0.22122034	14	−0.57226689
19	0.54162193	−0.31940281	0.27580153	−0.22773738	0.26665743	15	0.16037561
20	0.53311153	−0.32439488	0.24901558	−0.25270715	0.20903266	16	−0.75812475
						17	0.13790659
						18	−1.11141816
						19	0.10218995
						20	−2.42541817

not too small values of the time parameter $\nu t/a^2$, since the magnitude of $e^{-\varepsilon_m^2 \nu t/a^2}$ decreases very rapidly with increasing m. In the numerical evaluation of the previous results the coefficients C_m^* are based on $M = 20$, while fifteen terms of the series have been carried along in the determination of the velocity distribution. With this result the radial and axial velocity distribution could be determined. They are presented in Figures 3 and 4 for various time parameters $\nu t/a^2$ and ratios $u_0/U_0 = 0, 0.1, 0.5$ and 2.0, in all cases the flow volume per time unit $\dot{V}_0/2\pi a^2 U_0 = 1$. In Figure 3a the radial velocity distribution u/U_0 is presented for $u_0/U_0 = 0$ for various z-locations $\left(\frac{z}{a} = 0, 0.3, 0.6\right)$ and for the time $\nu t/a^2 = 0.005, 0.01, 0.02, 0.04$ and $\infty (\ldots)$. This is the case, where for the time region $t < 0$ we observe no suction across the wall. It may be seen that the steady flow case (... for $\nu t/a^2 = \infty$) is obtained in a rapid manner. The time to reach this case is indirectly proportional to the kinematic viscosity. For larger viscosity the transient time exhibits very small values. For a sudden change of the suction velocity from $u_0 = 0.1 U_0$ to U_0 at $t \doteq 0$ the radial velocity ratio u/U_0 is shown [7]. The radial velocity distribution during the time period $t < 0$ is given by the — . — curve, where the maximum radial velocity occurs at $r/a = \sqrt{2/3}$ and has the magnitude $u/U_0 = \frac{4}{3}\sqrt{\frac{2}{3}}\,(u_0/U_0)$. The sudden change of the suction velocity u at the wall $r = a$ results in a sudden jump of the radial velocity from $0.1 U_0$ to U_0 at $t = 0$. Again it may be noticed that the final profile for $\nu t/a^2 \to \infty$ is reached from below in a relatively rapid way for a comparatively small time parameter. For the cases $u_0/U_0 = 0.5$, 0.8, 2.0, 4.0 and 5.0 the radial velocity distribution is presented in Figure 3b, 3c and [7]. As in all cases the radial velocity is independent of the axial coordinate and shows for all z-values the same profile. In the case of $u_0/U_0 > 1.0$ (Figure 3c) and [7] the final velocity distribution exhibits a smaller profile than the original one at $t < 0$ and is reached from the top. The maximum of the radial velocity shifts during the transient in a decreasing fashion from a location closer to the axis of the tube towards the radial direction until it reaches the value $\frac{r}{a} = \sqrt{\frac{2}{3}}$. The axial velocity distribution is presented for $\dot{V}_0/2\pi a^2 U_0 = 1.0$ for various z-locations ($z/a = 0.03$ and 0.6), various dimensionless time parameters $\nu t/a^2 = 0.01$, 0.02, 0.04 and ∞ for the parameters $u_0/U_0 = 0, 0.1, 0.5, 0.8, 2.0, 4.0$ and 5.0 in Figures 4 a, b, c and [7]. The axial velocity distribution for the time region $t < 0$ is again given in form of the — . — curves. The velocity w changes with the axial coordinate and shows with increasing z-value a decreasing profile magnitude. In the case of $u_0/U_0 = 0$ the axial velocity profile w is unchanged in z-direction and exhibited by the — . — curve in Figure 4a. The sudden suction at the wall of the tube $r = a$ at the time $t = 0$ deforms the profile suddenly in such a way that near the wall the velocity w deviates from that for the time $t < 0$ to larger values, while near the center of the tube the axial velocity decreases considerably. The profile reaches during this transient phase with increasing time parameter $\nu t/a^2$ its final parabolic value for $\nu t/a^2 \to \infty$ with relatively small increases of the time parameter. Along the tube length the overall magnitude of the profile decreases linearly in z according to the constant suction at the wall $r = a$. It may be emphasized again, that at the inlet $z = 0$ a constant flow volume per time enters the pipeline, and that due to that the sudden magnitude change at the beginning of the suction at $t = 0$ is explained. For the suction ratio $u_0/U_0 = 0.1$ the axial velocity distribution across and along the pipeline may be seen in [7], where a similar behavior may be detected. Figure 3b exhibits the axial velocity w for $u_0 = 0.5 U_0$, i.e. for a case where the suction at the wall $r = a$ of the tube is doubled at $t = 0$ from a value $\frac{1}{2} U_0$ to U_0. The same may be seen in Figures of [7] for the suction ratio $u_0/U_0 = 0.8$, where again the velocity profiles drop during the transient below the final profile, which is reached, however, in a relatively small increase of the time parameter $\nu t/a^2$. In the case of a sudden decrease of the suction velocity at the wall $r = a$ of the tube, the axial velocity profile shows a different behavior, as may be seen in some Figures of [7]. Since at the inlet $z = 0$ the same flow volume enters the tube, the sudden decrease of the suction velocity at the wall $r = a$ of the tube increases (due to the pushing liquid from the inlet) the profile in the center of the tube, while near the wall of the tube a small decrease may be observed. With increasing transient time para-

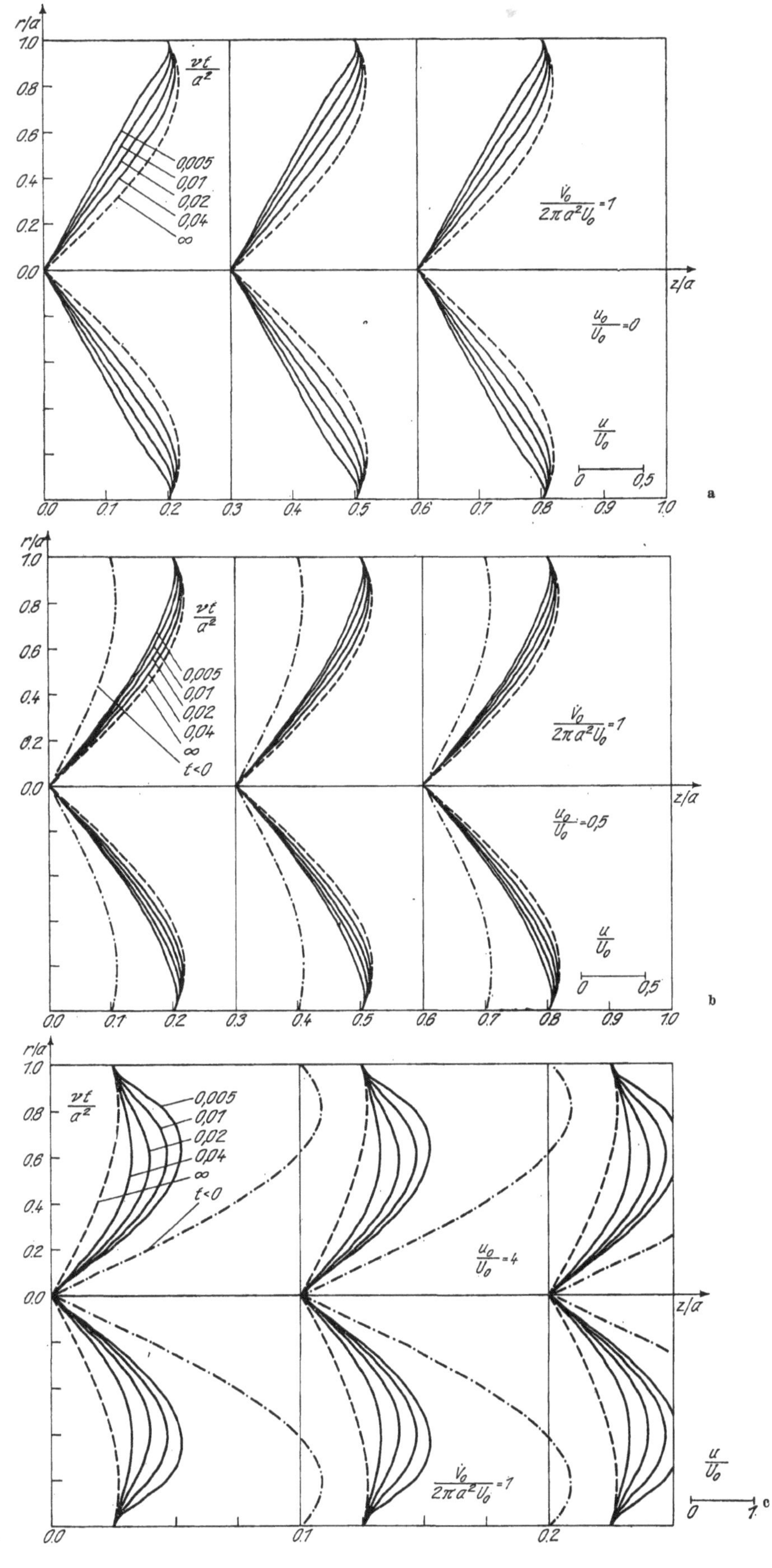

Fig. 3. Radial velocity distribution u

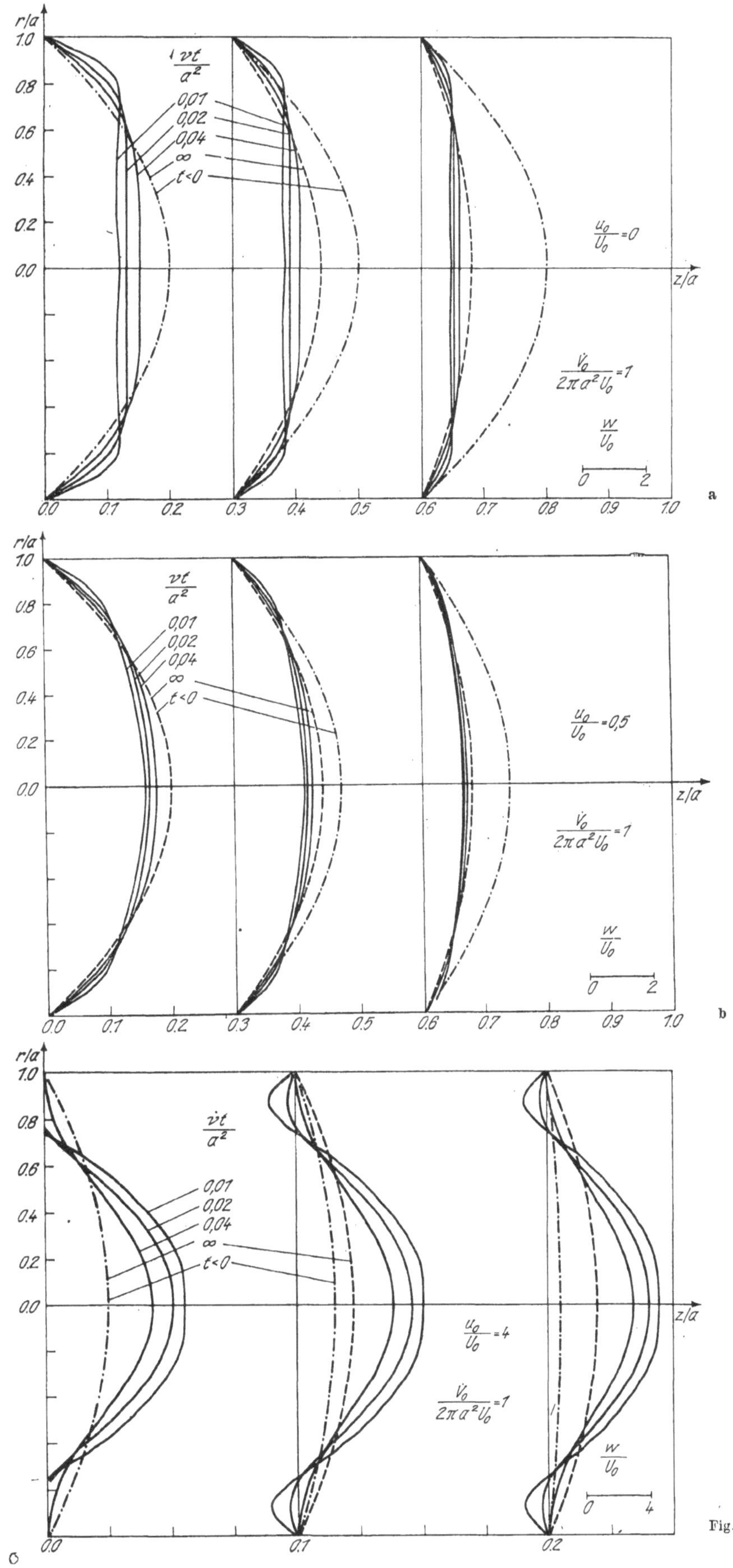

Fig. 4. Axial velocity distribution w

meter $\nu t/a^2$ the profile reduces its magnitude to that of the steady final case for $\nu t/a^2 \to \infty$ and assumes again parabolic shape. For an even larger decrease of the suction velocity $u_0 = 4U_0$ and $5U_0$, even near the wall a back flow region is observed, where the direction of the axial velocity is reversed, i.e., where the liquid flows in the opposite direction.

List of Symbols

a	radius of tube
C_m^*	constants (see Table)
J_m	Besselfunction of first kind and m^{th} order
p	liquid pressure
r, θ, z	polar coordinates
t	time
u_0, U_0	suction velocity at tube wall $r = a$
u	radial velocity
$\dot{V}_0$	flow volume/time unit
w	axial velocity
Γ	Gamma function
$\nu = \eta/\varrho$	kinematic viscosity
η	dynamic viscosity
ϱ	liquid density
Ψ	stream function
λ	$2\pi a u_0/\dot{V}_0$
ε_m	roots of $J_2(\varepsilon_m) = 0$
$\sigma_r, \sigma_\Theta, \sigma_z$	normal stresses
$\tau_{rz}, \tau_{r\Theta}, \tau_{\Theta z}$	shear stresses

References

1 Bauer, H. F., Strömung und Resorption im renalen Tubulus, Biomed. Technik **21**, (1976), S. 10–16.
2 Yuan, S. W.; Finkelstein, A. B., Laminar Pipe Flow with Injection and Suction through a porous Wall, ASME **78** (1956), pp. 719–724.
3 Peng, Y.; Yuan, S. W., Laminar Pipe Flow With Mass Transfer Cooling, J. Heat Transfer, Transact. ASME **87** (1965) pp. 252–258.
4 Bauer, H. F., Pulsatile Flow in a Straight Circular Pipe with Reabsorption across the Wall, Ing. Arch. **45**, (1976), S. 1–15.
5 Abramowitz, M.; Stegun, I. A. (ed.)., Handbook of Mathematical Functions, Dover Publications, Inc. New York 1965.
6 Luke, Y. L., Integrals of Besselfunctions, Mc Graw-Hill Book Company, Inc., New York 1962.
7 Bauer, H. F., Flow in a Cylindrical Tube with sudden Change of Suction across the wall, Forschungsbericht Hochschule der Bundeswehr München, LRT-WE-9-FB-16-1981.

Eingereicht am 27. 11. 1981

Anschrift: Prof. Dr. Helmut F. Bauer, Hochschule der Bundeswehr München, Institut für Raumfahrttechnik, Werner-Heisenberg-Weg 39, D 8014 Neubiberg, BRD

ZAMM 62, 521–531 (1982)

M. TESSAROTTO

Theoretical Features of the Radial Collisional Transport in Magnetoplasmas

Eine Formulierung des Stoß-Transport-Problems in einem nicht-symmetrisch konfigurierten Magnetoplasma wird erläutert. Es werden allgemeine Eigenschaften des Transports durch eine Familie von isobaren Flächen diskutiert, und es ergibt sich, daß Ambipolarität für den Partikelfluß nur für symmetrische Konfigurationen (d. h. mit einer vernachlässigbaren Koordinate) bis zur zweiten Ordnung im Larmor-Radius für ein quasineutrales Plasma automatisch erfüllt ist. Darüberhinaus werden im Fall eines Zustands nahe des thermodynamischen Gleichgewichts Integralausdrücke für den Fluß erhalten, und es wird eine partiell iterative Methode in Form der Drift-Fokker-Planck-Gleichung diskutiert. Schließlich werden die Grenzen des vorliegenden Modells untersucht, und es wird eine stochastische Interpretation der Ergebnisse in Form eines Irrfahrtmodells angegeben.

A formulation of the problem of collisional transport in magnetoplasmas in non-symmetric configurations is illustrated. General properties of the transport across a family of isobaric surfaces are discussed and it is found that only for symmetric configurations (e.g. with one ignorable coordinate) ambipolarity for the particle fluxes is automatically fulfilled, up to second order in the Larmor radius, for a quasi-neutral plasma. Furthermore integral expressions for the fluxes are obtained in the case of a state of near thermodynamic equilibrium and a partial iterative method in terms of the drift Fokker-Planck equation is discussed. Finally the limits of validity of the present model are investigated and a stochastic interpretation of the results in terms of the random-walk model is pointed out.

Обсуждается формулировка задачи ударного транспорта в несимметричной магнитной плазме. Анализируются общие свойства транспорта через семейство изобарических поверхностей и найдено, что амбиполярность для течения частиц выполняется автоматически только для симметричных конфигураций (т. е. с одной пренебрегаемой координатой) до второго порядка радиуса Лармора для квазинейтральной плазмы. Кроме того, для состояния вблизи термодинамического равновесия для течения получены интегральные выражения и обсуждён частично итерационный метод в виде уравнения Фоккера-Планка. Исследуются границы применения имеющейся модели и даётся стохастическая интерпретация результатов в виде модели блуждания.

1. Introduction

The theory of collision al transport in magnetoplasmas is a classical example in which kinetic effects may play a dominant role in the transport dynamics. The topic, although exhaustively investigated for toroidal axisymmetric configurations, appears still open in the case of non-symmetric magnetic equilibria. In such a case, in fact, the transport may be influenced not only by the diamagnetic drift motions as for the axisymmetric configurations, but also, in a complicate fashion, by the effect of non-conservation of the parallel adiabatic invariant on a flux surface, eventually giving rise to a secular drift motion of the particle orbits across many flux surfaces [1, 2].

This problem, on the other side, results of interest not only for the study of laboratory plasmas but also of relevance for the investigation of collisional processes in quiescent space plasmas. Therefore it encompasses a broad range of possible systems to which such a theory should eventually apply.

With the exception of [1] where the problem was formulated for a class of bounded sheared toroidal equilibrium configurations, most of the previous investigations have concerned the study of asymptotically small deviations ("ripples") from a toroidal axisymmetric configuration and were based on the use of an approximate model for the collision operator [2–7]. In [1] furthermore the drift FOKKER-PLANCK kinetic equation was solved introducing a further assumption on the distribution function — e.g. by taking it constant on a flux surface, in the case of a weakly collisional plasma — which in general needs not to be fulfilled. Elsewhere [8] a detailed account has been given for a formal kinetic theory of collisional transport for arbitrary bounded domains (in 3 dimensions), which has been applied to the investigation of a class of weakly collisional plasmas exhibiting a small relative variation of the magnetic field B along a flux (isobaric) surface (e.g. $|1 - B/B_{\max}| \ll 1$, with $B_{\max}$ the absolute maximum of B on that surface).

Purpose of this paper is to discuss certain general consequences which can be drawn from such a theory without having actually to solve the F — P kinetic equation explicitly or to introduce particular assumptions on the equilibrium configuration, apart obviously from those which assure the existence of the solution and the convergence of the perturbative expansion in terms of the LARMOR radius. More precisely, we introduce initially an appropriate "small LARMOR radius ordering" (SLRO) and the assumption of local near-thermodynamics equilibrium in order to limit the consideration to a regime of linear diffusion and to hydromagnetic equilibria which exhibit magnetic surfaces.

We formulate, furthermore, the problem of determination of the radial fluxes across a family of nested and closed surfaces by identifying them with the isobaric surfaces (with respect to which, the parallel transport does not contribute) and point out how, using an integral relationship obtained in terms of the F — P equation the fluxes may be expressed and in part exactly evaluated. In particular we are then able to prove in whole generality, up to second order in the LARMOR radius expansion, that certain contributions of the particle flux are always strictly ambipolar, a conclusion which is reached retaining the full F — P collision operator provided the plasma is assumed quasi-neutral at equilibrium. We are then able to determine for the "geometrical" contribution to the particle flux an exact integral expression by invoking simply the theorem of conservation of total linear momentum during a collision process. This result not only may facilitate the numerical computation of the (particle) flux but enables us to introduce a further exact iteration in terms of the drift F — P equation and extend the procedure to the consideration of the geometrical flux for an arbitrary dynamical variable $F_s(|\boldsymbol{v}|)$. In particular we discuss briefly also the case of symmetric configurations with one ignorable coordinate and show how in such a case the property of strict ambipolarity for the particle fluxes is authomatically fulfilled up to second order in the LARMOR radius for a quasineutral plasma.

The plan of the paper is as follows. In Section 2 the kinetic model is briefly recalled and the main assumptions are mentioned. In Section 3 the macroscopic equation for an arbitrary dynamical variable $F_s(v)$ is written and a general expression is obtained for the "radial" flux. In Section 4, after decomposing appropriately the flux, the "geometrical" contribution is examined in detail and its properties are discussed. In particular, it is shown how, proceeding to partial $\boldsymbol{v}$-integrations, it is possible to reduce the complicate integro-differential operator — acting on the unknown distribution function — to a simpler integral operator. The result is relevant in that it allows a further exact iteration in terms of the drift Fokker-Planck equation and a simple interpretation of the various flux contributions.

2. Basic Assumptions

We introduce the FOKKER-PLANCK kinetic equation for the distribution $f_s(\boldsymbol{r}, \boldsymbol{v}, t)$ of the species s-th ($s \in J$, the ensemble of the plasma species):

$$\left(\frac{\partial}{\partial t} + \boldsymbol{v} \cdot \nabla + \frac{e_s}{m_s}\left\{\boldsymbol{E} + \frac{\boldsymbol{v}}{c} \times \boldsymbol{B}\right\} \cdot \frac{\partial}{\partial \boldsymbol{v}}\right) f_s(\boldsymbol{r}, \boldsymbol{v}, t) = C_s(f \mid f) \tag{1}$$

where $C_s(f \mid f)$ is the F — P collision operator in the LANDAU form:

$$C_s(f \mid f) = \sum_k q_{sk} \frac{\partial}{\partial \boldsymbol{v}} \cdot \int d^3v' \frac{\partial^2 u}{\partial \boldsymbol{v}\, \partial \boldsymbol{v}} \cdot \left\{\frac{\partial}{\partial \boldsymbol{v}} f_s f_k(\boldsymbol{v}') - \frac{m_s}{m_k} \frac{\partial}{\partial \boldsymbol{v}'} f_k f_s(\boldsymbol{v})\right\} \tag{1'}$$

where

$$q_{sk} = 2\pi e_s^2 e_k^2 \ln \Lambda / m_s^2 \tag{1''}$$

and $\boldsymbol{E}, \boldsymbol{B}$ are respectively the electric and magnetic field obeying the MAXWELL's equation and e_s, m_s are the electric charge and mass of the s-th species. We consider a class of solutions $(f_s, \boldsymbol{E}, \boldsymbol{B})$ of such equations in the domain $\Omega \cup \delta\Omega \subseteq R^3$ (with $\delta\Omega$ boundary of Ω) and in the time interval $T = (t_0, t_1)$ for assigned initial and boundary conditions. More precisely, we intend to limit the class of possible initial and boundary conditions by introducing an appropriate set of assumptions which are chosen: a) to limit the class of possible hydromagnetic equilibria to configurations which admit at most a countable set of nested magnetic surfaces (e.g. isobaric surfaces); b) to consider a regime of linear diffusion from a near-equilibrium state. In order to specify further the class of mechanical systems it is convenient to look at the physically interesting time scales which result from eq. (1). They result, respectively, the LARMOR frequency $\Omega_s = e_s B / m_s c$, the bounce frequency along a magnetic flux line $\omega_b = \left(\oint_C ds/|v_{||}|\right)^{-1}$ (with C an unperturbed particle orbit and $v_{||} = \boldsymbol{v} \cdot \hat{\boldsymbol{n}}$ where $\hat{\boldsymbol{n}} = \boldsymbol{B}/B$), the drift-bounce frequency with which the guiding center drifts around the system $\omega_{Db} = \left\{L \oint_C ds v_D/|v_{||}| / \oint_C ds/|v_{||}|\right\}^{-1}$ with velocity $\boldsymbol{v}_D$ and an effective collision frequency ν_s, to be appropriately defined.

Denoting with $r_{l,s} = v_{\text{th},s}/\Omega_s$ the LARMOR radius and L a scale length representing the linear dimension of the system (if it is bounded) or a typical length characterizing the inhomogeneities of the equilibrium configuration, the previous frequencies result in the ratios:

$$\frac{\omega_b}{\Omega_s} \sim \varepsilon_s; \qquad \frac{\omega_{Db}}{\Omega_s} \lesssim \varepsilon_s^2 \quad \text{with} \quad \varepsilon_s = r_{l,s}/L \tag{2}$$

whereas ν_s/Ω_s remains arbitrary. Since in the sequel we are furthermore interested in studying the effect of COULOMB collisions on diffusion we shall require:

$$\frac{\nu_s}{\Omega_s} \sim \varepsilon_s \tag{3}$$

and introduce a perturbative expansion in terms of the ratio ε_s, assumed $\ll 1$. This is the so-called small LARMOR radius ordering (SLRO) (see for example ref. [8] for further discussions) which can be expressed formally by requesting that at time $t = t_0$ a solution $(f_s, \boldsymbol{E}, \boldsymbol{B})$ exist in $\Omega \cup \delta\Omega$ (and exibits all the appropriate regularity properties necessary for the subsequent developments) and fulfills the asymptotic ordering:

$$\frac{\nu_s}{\Omega_s} \sim \frac{c|\nabla\Phi|}{v_{\text{th},s} B} \sim \left(\frac{cE^{\text{rot}}}{v_{\text{th},s} B}\right)^{1/2} \sim O(\varepsilon_s) \ll 1\,, \qquad \frac{\omega_{pe}^2}{\Omega_e^2} \sim \beta_0 \sim O(\varepsilon_s)^0; \tag{4}$$

Here $\boldsymbol{E}^{\text{rot}}$ is the rotational part of the electric field, for which the condition of consistency on the parallel current $\partial j_{||}/\partial t = 0$ must be fulfilled, since from the third MAXWELL equation a time-varying electric current determines a rotational electric field two orders of magnitude (in terms of $O(\varepsilon_s)$) larger. Furthermore Φ is the electrostatic potential, β_0 the ratio between kinetic and magnetic pressure, which in the context of the present theory is assumed $\ll 1$ but possibly not small in ε_s, and ω_{pe} is the electron-plasma frequency which is scaled in order to assure the quasineutrality of the equilibrium configuration.

After having expanded each physical quantity of relevance in power series of ε_s, to first order one obtains then:

$$f_{1,s} = \tilde{f}_{1,s} + \bar{f}_{1,s}\,, \tag{5}$$

where

$$\tilde{f}_{1,s} = \frac{\boldsymbol{v} \times \boldsymbol{n}}{\Omega_s} \cdot \nabla f_{0,s} \tag{6}$$

and $\bar{f}_{1,s}$ is determined by the drift F — P equation:

$$v_{||}\boldsymbol{n} \cdot \nabla \bar{f}_{1,s} = -\boldsymbol{v}_{D,s} \cdot \nabla f_{0,s} + \frac{e_s}{T_{0,s}} v_{||} E_{||}^{\text{rot}} f_{0,s} + C_s(f_0 \mid \bar{f}_1) \tag{7}$$

where $\boldsymbol{v}_{D,s}$ is the diamagnetic drift velocity:

$$\boldsymbol{v}_{D,s} = \frac{\boldsymbol{n}}{\Omega_s} \times (\mu \nabla B + v_{||}^2 \boldsymbol{n} \cdot \nabla \boldsymbol{n})\,. \tag{8}$$

Denoting, for simplicity, $\boldsymbol{B}$ the $O((\varepsilon_s)^0)$-order contribution to the magnetic field, it follows thanks to AMPERE's law and neglecting higher-order inductive terms:

$$(\nabla \times \boldsymbol{B}) \times \boldsymbol{B} = 4\pi \nabla p_0 \tag{9}$$

e.g. the hydromagnetic equilibrium equation for $\boldsymbol{B}$. Here $p_0 = \sum_{s\text{This}} N_{0,s} T_{0,s}$ is the leading-order contribution to the scalar pressure. This proves consequently that SLRO together with the assumption of near-thermodynamic equilibrium (which implies for $f_{0,s}$ the choice of a local maxwellian distribution), automatically imposes the fluid constraint (9). Then, eq. (9) is a necessary condition for the existence of solutions for eq. (1), in validity of SLRO.

Assuming therefore that the class of mechanical systems fulfilling such hypotheses be non-empty, and that in a time interval $T = (t_0, t_1)$ the perturbative solution $(f_s, \boldsymbol{E}, \boldsymbol{B})$ converges, we intend to proceed to the precise formulation of the problem of the derivation of the "radial" fluxes for a dynamical variable F_s.

3. Derivation of the Radial Fluxes

In order to obtain a macroscopic description of the mechanical system, a problem which shall be further examined elsewhere (ref. [10]), we do not need, at least in principle, the full F — P equation or even the equations of the perturbative expansion previously introduced. In fact what is sufficient involves much less than the precise knowledge of the distribution function f_s (and in terms of it the e.m. field), and requires the determination of certain appropriate moments of the distribution function itself. To this purpose we proceed, then, to derive from the F — P equation (1) an appropriate macroscopic equation for a dynamical variable $F_s(v)$ which shall later be identified with $F_s = 1$, yielding the equation of continuity or particle balance equation, and $F_s = m_s v^2/2$, which corresponds to the equation of energy balance. The macroscopic equation for $F_s(|\boldsymbol{v}|)$ results:

$$\xi \left\langle \int \mathrm{d}^3 v \left(F_s \frac{\partial}{\partial t} f_s - \frac{e_s}{m_s} f_s \boldsymbol{E} \cdot \frac{\partial}{\partial \boldsymbol{v}} F_s - F_s C_s(f \mid f) \right) \right\rangle - \frac{\partial}{\partial p_0} \Gamma_F^{(s)} = 0 \tag{10}$$

where we have introduced the surface-average:

$$\langle A \rangle = \xi^{-1} \iint_{\Sigma(p_0)} \mathrm{d}\sigma\, A/|\nabla p_0| \quad \text{with} \quad \xi = \iint_{\Sigma(p_0)} \mathrm{d}\sigma/|\nabla p_0| \tag{11}$$

with $\Sigma(p_0)$ an isobaric surface if $\Omega \cup \delta\Omega$ is bounded and $\nabla p_0 \neq 0$ (except, at most, a countable set of closed loops enclosed in $\Omega \cup \delta\Omega$). Furthermore we have assumed, for definiteness, that denoting with $V(p_0)$ the domain enclosed by $\Sigma(p_0)$, p_0 decreases outward with respect to $V(p_0)$ and for the further discussion it is irrelevant whether $\Omega \cup \delta\Omega$ is covered by one or more families of nested isobaric surfaces (each corresponding to a magnetic axis on which the pressure is maximum, e. g. $\nabla p_0 = 0$). In eq. (10) $\Gamma_F^{(s)}$ denotes the flux of the dynamical variable F_s (across the isobaric surface $\Sigma(p_0)$):

$$\Gamma_F^{(s)} = \xi \langle \nabla p_0 \cdot \int \mathrm{d}^3 v\, \boldsymbol{v} F_s f_s \rangle\,. \tag{12}$$

In the case of an unbounded domain $\Omega \cup \delta\Omega$, in order to assure the existence of the integrals appearing in the macroscopic equation, one simply needs to cut the integration domain $\Sigma(p_0)$ with two arbitrary regular surfaces Σ_1 and Σ_2, which may be chosen appropriately, and add in correspondence on the r.h.s. of eq. (10) the further flux contributions:

$$\frac{\partial}{\partial p_0} \left(\iint_{\Sigma} \mathrm{d}\sigma_1\, \boldsymbol{n}_1 + \iint_{\Sigma} \mathrm{d}\sigma_2 \boldsymbol{v}_2 \right) \cdot \int \mathrm{d}^3 v\, \boldsymbol{v} F_s f_s{}' \tag{10'}$$

where $\boldsymbol{n}_1$ and $\boldsymbol{n}_2$ are the unit vectors normal to Σ_1 and Σ_2, directed "outward" with respect to the domain $V(p_0)$. Therefore the two cases, of bounded and unbounded domain $\Omega \cup \delta\Omega$, can be treated much in the same way. In the sequel, however, we shall report for simplicity only the results pertaining to the first case.

Integrating by parts in $\boldsymbol{v}$-space with respect to the azimuth θ (here we use as $\boldsymbol{v}$-space coordinates $E_s = v^2/2 - e_s\Phi_0/m_s$, $\mu = v_\perp^2/2B$, $\theta = \text{arctg}\,(\boldsymbol{v}\cdot\hat{b}/\boldsymbol{v}\cdot\hat{p})$ with $(\hat{n}, \hat{b}, \hat{p})$ a right-handed set of orthogonal unit vectors and Φ_0 the $O(\varepsilon_s)_0$ part of the e.s. potential) and invoking the second order (e.g. $O(\varepsilon_s)^2$) F — P equation, not reported here for brevity, we obtain:

$$\Gamma_F^{(s)} = \Gamma_{FG}^{(s)} + \Gamma_{FE}^{(s)} + \Gamma_{FC}^{(s)} \tag{13}$$

where $\Gamma_{FG}^{(s)}$, which we propose to identify as "geometrical" flux (and coincides with the so-called toroidal contribution for toroidal axisymmetric configurations), the electric-drift $\Gamma_{FE}^{(s)}$ and classical (or SPITZER's) $\Gamma_{FC}^{(s)}$ fluxes read respectively:

$$\Gamma_{FG}^{(s)} = \xi\left\langle \Omega_s^{-1}\nabla p_0\cdot\int \mathrm{d}^3v\,\boldsymbol{v}\times\boldsymbol{n}F_s\left(\boldsymbol{v}_\perp\cdot\nabla(f_{0,s}+\bar{f}_{1,s}) - \boldsymbol{v}_\perp\cdot(\mu\nabla B + v_\parallel^2\boldsymbol{n}\cdot\nabla\boldsymbol{n})\frac{\partial}{B\,\partial\mu}\bar{f}_{1,s}\right)\right\rangle, \tag{14}$$

$$\Gamma_{FE}^{(s)} = -\xi\left\langle \Omega_s^{-1}\nabla p_0\cdot\int \mathrm{d}^3v\,\boldsymbol{v}\times\boldsymbol{n}F_s\frac{e_s}{T_{0,s}}\boldsymbol{E}^{\text{rot}}\cdot\boldsymbol{v}_\perp f_{0,s}\right\rangle, \tag{15}$$

$$\Gamma_{FC}^{(s)} = \xi\Omega_s^{-1}\nabla p_0\cdot\int \mathrm{d}^3v\,\boldsymbol{v}\times\boldsymbol{n}F_sC_s(f_0\,|\,\tilde{f}_1)\rangle\,. \tag{16}$$

After straightforward integrations by parts in E_s, μ and $\hat{\boldsymbol{n}}\cdot\nabla$ the geometrical and electric drift fluxes read:

$$\Gamma_{FG}^{(s)} = \xi\langle\nabla p_0\cdot\int \mathrm{d}^3v\,\boldsymbol{v}_{D,s}F_s\bar{f}_{1,s}\rangle - \xi\langle\Omega_s^{-1}\nabla p_0\cdot\int \mathrm{d}^3v\,\boldsymbol{v}\times\boldsymbol{n}F_s\boldsymbol{v}_\perp\cdot\nabla f_{0,s}\rangle\,, \tag{14'}$$

$$\Gamma_{FE}^{(s)} = -\frac{c}{T_{0,s}}\xi\left\langle B^{-1}\boldsymbol{n}\times\nabla p_0\cdot\boldsymbol{E}^{\text{rot}}\int \mathrm{d}^3v\,\frac{v_\perp^2}{2}F_sf_{0,s}\right\rangle. \tag{15'}$$

The expressions of $\Gamma_{FE}^{(s)}$ and $\Gamma_{FC}^{(s)}$ can be worked out immediately only for the former term. After tedious calculations it is possible to determine exactly, to leading order, also the classical contribution $\Gamma_{FC}^{(s)}$ (the result is reported in [3] for $F_s = 1$, $m_sv^2/2$).

We note, furthermore, that on the r.h.s. of eq. (14') we have included a contribution due to $f_{0,s}$ itself which may be ignored for configurations with shear (at least in validity of the assumption $\nu_s/\Omega_s \sim O(\varepsilon_s)$). We intend now to show how further information may be obtained on $\Gamma_{FG}^{(s)}$ by introducing an iteration in terms of the drift equation (7).

4. Properties of the Geometrical Flux

For the subsequent developments we find convenient to use as $\boldsymbol{v}$-space variables the set v, $\lambda = 2\mu/v^2$ and θ, in which case introducing the jacobian of the transformation $(v_x, v_y, v_z) \to (v, \lambda, \theta)$ results:

$$\int \mathrm{d}^3v = \frac{1}{2}\sum_{\sigma=\pm1}\oint \mathrm{d}\theta\int_0^\infty \mathrm{d}v v^2\int_0^{1/B}\mathrm{d}\lambda\,B/(1-\lambda B)^{1/2} \quad \text{with} \quad \sigma = \text{sign}\,(v_\parallel)\,. \tag{17}$$

Using the definition (8) to write:

$$\nabla p_0\cdot\boldsymbol{v}_{D,s} = -\frac{v_\parallel}{\Omega_s}\nabla p_0\cdot\boldsymbol{B}\times\nabla\left(\frac{v_\parallel}{B}\right) \tag{18}$$

we obtain for $\Gamma_{FG}^{(s)}$:

$$\Gamma_{FG}^{(s)} = \frac{m_s\pi}{e_s}\sum_{\sigma=\pm1}\int_0^\infty dvv^3\xi\left(\int_0^{1/B}\mathrm{d}\lambda\,F_s\bar{f}_{1,s}\boldsymbol{B}\cdot\nabla g(v,\lambda,\sigma,s,x,y)\right) \tag{19}$$

where $g(v, \lambda, \sigma, s, x, y)$ is a solution of the equation:

$$\boldsymbol{n}\cdot\nabla g(v,\lambda,\sigma,s,x,y) = -\nabla p_0\cdot\boldsymbol{n}\times\nabla\left(\frac{v_\parallel}{B}\right) \tag{20}$$

with $g(v, \lambda, \sigma, s_0, x, y) = 0$ for $s = s_0$ such that $B(s_0, x, y) = B_{\min}$ the absolute maximum of B along a closed field line C and we have introduced, an orthogonal right-handed system of coordinate (s, x, y) (see Appendix A). Therefore results:

$$g(v,\lambda,\sigma,s,x,y) = \int_{s_0}^{s}\mathrm{d}s'\,\nabla p_0\cdot\boldsymbol{n}\times\nabla\left(\frac{v_\parallel}{B}\right). \tag{21}$$

In order to analyze the structure of $\Gamma_{FG}^{(s)}$ it is convenient to proceed to a partial iteration in terms of the drift equation (7). For this purpose we invoke eq. (B. 8) (see Appendix B) and eq. (7) to iterate in terms of $\boldsymbol{n}\cdot\nabla\bar{f}_{1,s}$. We obtain then:

$$\Gamma_{FG}^{(s)} = \Lambda_F^{(s)} + \Pi_{FE}^{(s)} + \Pi_F^{(s)} \tag{22}$$

with the positions:

$$\Lambda_F^{(s)} = \frac{2\pi m_s c}{e_s}\int_0^\infty dv\, v^3 F_s \xi \left\langle\left(B\int_0^{1/B_{\max}} d\lambda\, P\bar{f}_{1,s}[g(v,\lambda,\sigma,s,x,y)]_{s=s_1}\delta(s-s_1) - \right.\right.$$

$$\left.\left. - \nabla\cdot\boldsymbol{n}\{P\bar{f}_{1,s}g(v,\lambda,\sigma,s,x,y)\}_{\lambda=1/B}\right)\right\rangle, \tag{23}$$

$$\Pi_{FE}^{(s)} = \frac{m_s c}{T_{0,s}}\xi\left\langle E_{\parallel}^{\rm rot}\int d^3v\, F_s f_{0,s} g(v,\lambda,\sigma,s,x,y)\right\rangle, \qquad \Pi_F^{(s)} = \frac{m_s c}{e_s}\xi\langle I_F^{(s)}\rangle \tag{24), (25}$$

and

$$I_F^{(s)} = \int d^3v\, F_s g(v,\lambda,\sigma,s,x,y)\, C_s(f_0 \mid D\bar{f}_1). \tag{26}$$

where $P\bar{f}_{1,s}$ and $D\bar{f}_{1,s}$ are respectively the even and odd parts of $\bar{f}_{1,s}$ in terms of $v_{\parallel}$. Note, furthermore, that $[g(v,\lambda,\sigma,s,x,y)]_{s=s_1}$ is the jump:

$$[g(v,\lambda,\sigma,s,x,y)]_{s=s_1} = g(v,\lambda,\sigma,s_1+s(C),x,y) - g(v,\lambda,\sigma,s_1,x,y) \tag{27}$$

along a closed field line after one turn starting from the point (s_1, x, y) $\big(s_1 + s(C)$ defines then the determination of s after one turn$\big)$.

The contribution $\Lambda_F^{(s)}$ can be given a more explicit expression by using eq. (B. 7). Eq. (23) delivers then:

$$\Lambda_F^{(s)} = \frac{2\pi m_s c}{e_s}\int_0^\infty dv\, v^3 F_s \xi \left\langle\left(\int_0^{1/B_{\max}} d\lambda\, P\bar{f}_{1,s}[g(v,\lambda,\sigma,s,x,y)]_{s=s_1}\delta(s-s_1) + \right.\right.$$

$$\left.\left. + \int_{1/B_{\max}}^{1/B_{\min}} d\lambda\, P\bar{f}_{1,s} g(v,\lambda,\sigma,s,x,y)\big(\delta(s-s_a(\lambda)) - \delta(s-s_b(\lambda))\big)\right)\right\rangle \tag{28}$$

where $B_{\min}$ is the absolute minimum of B along a closed field line. We note that eq. (28) follows by introducing for the second term on the r.h.s. of eq. (23) the transformation $ds\,\boldsymbol{n}\cdot\nabla(1/B) = d\lambda$, which restores the λ-integration. The two terms appearing on the r.h.s. of eq. (28) result, therefore, simply as the contributions due to circulating and trapped particles.

It is also instructive to express such integrals in terms of the parallel adiabatic invariant $J_{\parallel} = \oint ds\, v_{\parallel}$, by transforming from the coordinates (s, x, y) to (α, β, γ) (see Appendix A). In such a case results:

$$\oint_C ds\, \nabla p_0 \cdot \boldsymbol{n}\times\nabla\left(\frac{v_{\parallel}}{B}\right) = \frac{\partial}{\partial\beta}J_{\parallel} \tag{29}$$

where C denotes the zeroth-order unperturbed particle orbit and we have identified $\alpha = p_0$. Equation (28) then yields:

$$\Lambda_F^{(s)} = \frac{2\pi m_s c}{e_s}\int_0^\infty dv\, v^3 F_s \xi \left\langle B\left(\int_0^{1/B_{\max}} d\lambda\, P\bar{f}_{1,s}\frac{\partial}{\partial\beta}J_{\parallel}\delta(s-s_1) + \int_{1/B_{\max}}^{1/B_{\min}} d\lambda\, P\bar{f}_{1,s}\frac{\partial}{\partial\beta}J_{\parallel}\big(\delta(s-s_a(\lambda)) - \delta(s-s_b(\lambda))\big)\right)\right\rangle \tag{30}$$

if we assume that $P\bar{f}_{1,s}$ be symmetric with respect to the exchange of mirror points, e.g.:

$$P\bar{f}_{1,s}(v,\lambda,s_a(\lambda),x,y) = P\bar{f}_{1,s}(v,\lambda,s_b(\lambda),x,y)\,. \tag{31}$$

In order to exploit the contribution $\Pi_F^{(s)}$ (eq. (25)) it is convenient to express it as a linear integral operator acting on $D\bar{f}_{1,s}$. After repeated integrations by parts results:

$$I_F^{(s)} = \sum_k q_{sk}\int d^3v\,(D\bar{f}_{1,s}V_{ks} - D\bar{f}_{1,k}W_{sk}) \tag{32}$$

where

$$V_{ks} = \left(1 + \frac{m_s}{m_k}\right)\frac{\partial}{\partial\boldsymbol{v}}H_s\cdot\boldsymbol{C}_k + \frac{\partial^2}{\partial\boldsymbol{v}\,\partial\boldsymbol{v}}H_s : \mathbf{D}k\,, \tag{33}$$

$$W_{sk} = \left(1 + \frac{m_s}{m_k}\right)\int d^3v'\,\frac{\partial^2 u}{\partial\boldsymbol{v}\,\partial\boldsymbol{v}} : \frac{\partial}{\partial\boldsymbol{v}'}H_s\frac{\partial}{\partial\boldsymbol{v}'}f_{0,s} + \frac{m_s}{m_k}\int d^3v'\,\frac{\partial^2 u}{\partial\boldsymbol{v}\,\partial\boldsymbol{v}} : \frac{\partial^2}{\partial\boldsymbol{v}'\,\partial\boldsymbol{v}'}H_s f_{0,s} \tag{34}$$

and

$$\boldsymbol{C}_k = \frac{\partial}{\partial\boldsymbol{v}}\cdot\int d^3v'\,\frac{\partial^2 u}{\partial\boldsymbol{v}\,\partial\boldsymbol{v}}f_{0,k}\,, \qquad \mathbf{D}_k = \int d^3v'\,\frac{\partial^2 u}{\partial\boldsymbol{v}\,\partial\boldsymbol{v}}f_{0,k}\,, \qquad H_s = F_s g(v,\lambda,\sigma,s,x,y)\,. \tag{35), (36), (37}$$

By integrating eq.s (35) and (36) we obtain:

$$\boldsymbol{C}_k = 4\pi v_{\text{th},k}^2 \boldsymbol{v}/v \left(f_{0,k} - \frac{1}{v} \int_0^v \mathrm{d}v' f_{0,k} \right), \tag{38}$$

$$\mathbf{D}_k = 4\pi v_{\text{th},k}^2 f_{0,k} \left(\Pi \left(\frac{2}{3} x_k + \frac{1}{4x_k^2} - \frac{1}{3} \right) + \frac{\boldsymbol{vv}}{v^2} \left\{ \frac{2}{3} x_k (1 + x_k - 2x_k^2) - \frac{3}{4x_k^2} \right\} \right) +$$
$$+ 2\pi v_{\text{th},k}^2 \left(\Pi \left(1 - \frac{1}{2x_k^2} \right) - \frac{\boldsymbol{vv}}{v^2} \left(1 - \frac{3}{2x_k^2} \right) \right) \frac{1}{v} \int_0^v \mathrm{d}v' f_{0,k}. \tag{39}$$

It is clear that also for the term $\Pi_F^{(s)}$ (as before for $\Gamma_{FG}^{(s)}$) it is possible to proceed to a further (partial) iteration in terms of the drift eq. (7). This is achieved introducing the functions $G_{ks}(v, \lambda, \sigma, s, x, y)$ and $H_{sk}(v, \lambda, \sigma, s, x, y)$ solutions of the equations:

$$\boldsymbol{n} \cdot \nabla G_{ks} = V_{ks}, \qquad \boldsymbol{n} \cdot \nabla H_{sk} = W_{sk} \tag{40), (41}$$

with boundary conditions $G_{ks}(v, \lambda, \sigma, s_0, x, y) = H_{sk}(v, \lambda, \sigma, s_0, x, y) = 0$. Integrating by parts, noting that the contributions in the mirror points vanish, yields:

$$\Pi_F^{(s)} = \Pi_{FC}^{(s)} + \Pi_{FCD}^{(s)} + \Pi_{FCC}^{(s)} \tag{42}$$

with the positions:

$$\Pi_{FC}^{(s)} = \frac{m_s c}{2e_s} \sum_k q_{sk} \int_0^\infty \mathrm{d}v\, v^2 \sum_{\sigma=\pm 1} \xi \left\langle B \int_0^{1/B_{\max}} \mathrm{d}\lambda \left(\bar{D} f_{1,s} \left| G_{ks} \right|_{s=s_1} \delta(s - s_1) - \bar{D} f_{1,k} \left| H_{sk} \right|_{s=s_1} \delta(s - s_1) \right) \right\rangle, \tag{43}$$

$$\Pi_{FCD}^{(s)} = \frac{m_s c}{e_s} \sum_k q_{sk} \xi \left\langle \int \mathrm{d}^3 v \frac{1}{v} \{ \boldsymbol{v}_{D,s} \cdot \nabla f_{0,s} G_{ks} - \boldsymbol{v}_{D,k} \cdot \nabla f_{0,k} H_{sk} \} \right\rangle, \tag{44}$$

$$\Pi_{FCC}^{(s)} = -\frac{m_s c}{e_s} \sum_k q_{sk} \xi \left\langle \int \mathrm{d}^3 v \frac{1}{v} \{ C_s(f_0 \mid P\bar{f}_1)\, G_{ks} - C_k(f_0 \mid P\bar{f}_1)\, H_{sk} \} \right\rangle. \tag{45}$$

The results obtained for $\Gamma_{FG}^{(s)}$ may be collected by writing:

$$\Gamma_{FG}^{(s)} = \Lambda_F^{(s)} + \Pi_{FE}^{(s)} + \Pi_{FC}^{(s)} + \Pi_{FCD}^{(s)} + \Pi_{FCC}^{(s)} \tag{46}$$

(in terms of eq.s (22) and (42)), and introducing the notations:

$$\Gamma_{FPS}^{(s)} = \Gamma_{FE}^{(s)} + \Gamma_{FC}^{(s)} + \Pi_{FE}^{(s)} + \Pi_{FCD}^{(s)}, \qquad \Gamma_{FRG}^{(s)} = \Lambda_F^{(s)} + \Pi_{FC}^{(s)} + \Pi_{FCC}^{(s)} \tag{47), (48}$$

with

$$\Gamma_F^{(s)} = \Gamma_{FPS}^{(s)} + \Gamma_{FRG}^{(s)}. \tag{49}$$

Here the flux contribution $\Gamma_{FPS}^{(s)}$ has been defined by collecting in it all the various terms which already result expressed by means of a functional of known quantities ($f_{0,s}, \bar{f}_{1,s}, \boldsymbol{E}^{\text{rot}}, \boldsymbol{B}$), whereas the 'residual' geometrical contributions — involving the still undetermined distribution $\bar{f}_{1,s}$ — appear in $\Gamma_{FRG}^{(s)}$.

Thus, in terms of such a decomposition we have reduced the problem to the determination of $\Gamma_{FRG}^{(s)}$ alone, a task which elsewhere shall be developed in detail for the case of axisymmetric toroidal plasmas [9]. Its utility, apart from the application to symmetric configurations with for which an illustration shall be given shortly, is the possibility of obtaining a physical interpretation of the various collisional transport processes in a magnetized plasma.

In particular, it is obvious the $\Lambda_F^{(s)}$ is produced by the non-conservation of the parallel adiabatic invariant on a prescribed isobaric surface, and is therefore associated to the particle drift motion across the family of isobaric surfaces. Elsewhere [8] its computation, for a weakly collisional plasma and for a magnetic field weakly varying on a given isobaric surface ($|1 - B/B_{\max}| \ll 1$), has been reported. In this limit $\Lambda_F^{(s)}$ yields the leading-order contribution in terms of a perturbative expansion in the ratio between the effective collision frequency and the bounce frequency along the unperturbed particle orbit. It is therefore also relevant for the investigation of the problem of toroidal 'ripples' (e.g. asymptotically small perturbations of an axisymmetric toroidal magnetic field $\boldsymbol{B}$) [3, 4, 6]. Note that a particular result which may be inferred for weakly collisional plasmas immersed in a sheared configuration is that in $\Lambda_F^{(s)}$ the dominant contribution should be expected from trapped particles. In this respect it is instructive to recall the wellknown results for a collisionless plasma [11]. In fact, even if $J_{||}$ is not conserved on a single unperturbed orbit of a circulating particle, it results nevertheless conserved in an appropriate average sense on the whole magnetic surface. On the other hand, if the magnetic flux lines envelop ergodically a given isobaric surface, it is implied the quasi-conservation of $J_{||}$ also on a given (unperturbed) particle orbit. The argument does not apply evidently to trapped particles and suggests that the contribution due to $\Lambda_F^{(s)}$ should be enhanced for configurations which exhibit strong variations of the magnetic field along a prescribed isobaric surface ($B_{\max} \gg B_{\min}$).

As far as concerns $\Gamma_{FPS}^{(s)}$, its computation is in principle possible without difficulties. We note, in particular, that the classical ($\Gamma_{FC}^{(s)}$) and electric-drift ($\Gamma_{FE}^{(s)}$) fluxes, correspond to the wellknown cylindrical contributions — reported elsewhere [8] for a multispecies plasma — which are the only ones surviving in the case of a (infinite) cylindrical and axially symmetric configuration. Moreover, $\Pi_{FE}^{(s)}$ and $\Pi_{FCD}^{(s)}$ can be interpreted, respectively as the 'geometrical' contribution due to $E_{\|}^{\mathrm{rot}}$ and the collisionless diamagnetic part of $\bar{f}_{1,s}$, appearing — together with $\Gamma_{FRG}^{(s)}$ — when the symmetry of the previous configuration is broken in some way.

In order to obtain further insight let us now cast an eye on the special class of hydromagnetic equilibria with one ignorable coordinate [13]. More precisely, introducing for convenience an orthogonal right-handed system of coordinates $(x_1 = p_0, x_2, x_3)$ with (x_2, x_3) curvilinear coordinates on a given isobaric surface $\Sigma(p_0)$ let us assume that $\boldsymbol{B}$ results invariant with respect to a 'translation' of the type:

$$x_3 \to x_3' = x_3 + k\,. \tag{50}$$

Since therefore x_3 is ignorable for all physical quantities of relevance for the system the surface average (11) becomes:

$$\langle A\rangle = \xi^{-1} \oint \mathrm{d}x_2\, A/(|\nabla x_2|\, |\nabla x_3|\, |\nabla p_0|) \tag{51}$$

with

$$\xi = \oint \mathrm{d}x_2/(|\nabla x_2|\, |\nabla x_3|\, |\nabla p_0|)\,. \tag{52}$$

Moreover, since we can write $\boldsymbol{B} = \hat{e}_2 B_2 + \hat{e}_3 B_3$, results $\boldsymbol{n}\cdot\nabla A = \frac{B_2}{B}|\nabla x_2|\,\frac{\partial}{\partial x_2}A$ and from eq. (18):

$$\nabla p_0 \cdot \boldsymbol{v}_{D,s} = \frac{v_{\|}}{\Omega_s}|\nabla p_0|\; B_3 |\nabla x_2|\,\frac{\partial}{\partial x_2}\left(\frac{v_{\|}}{B}\right). \tag{53}$$

From eq. (20) we obtain then:

$$g(v, \lambda, \sigma, p_0, x_2, x_3) = \frac{|\nabla p_0|\; B_3}{B_2}\,\frac{v_{\|}}{B}, \tag{54}$$

in fact from AMPERE's law we infer $\frac{\partial}{\partial x_2}\,\frac{|\nabla p_0|\, B_3}{B_2} = 0$.

An immediate consequence of eq. (54) is $\frac{\partial}{\partial x_2}J_{\|} = 0\left(\text{e.g. } \frac{\partial}{\partial\beta}J_{\|} = 0\right)$, namely $\Lambda_F^{(s)}$ and $\Pi_{FC}^{(s)}$ then vanish due to the conservation of the parallel adiabatic invariant on a given surface $\Sigma(p_0)$. Furthermore, the expressions of $\Pi_{FCD}^{(s)}$ and $\Pi_{FCC}^{(s)}$ result greatly simplified. Considering here for simplicity only the case of particle fluxes $(F_s = 1)$ we obtain then from eq. (37) $H_s = g$ (where g is given by eq. (54)), so that:

$$\frac{\partial}{\partial \boldsymbol{v}}H_s = \boldsymbol{n}\,\frac{|\nabla p_0|\; B_3}{BB_2} \quad\text{and}\quad \frac{\partial^2}{\partial \boldsymbol{v}\,\partial \boldsymbol{v}}H_s = 0 \tag{55}$$

and from eqs. (33) and (34):

$$V_{ks} = \left(1 + \frac{m_s}{m_k}\right)\boldsymbol{V}_k v_{\|}\,, \qquad W_{sk} = \left(1 + \frac{m_s}{m_k}\right)\boldsymbol{V}_s v_{\|} \tag{56), (57}$$

with

$$\boldsymbol{V}_k = x_k^{-2}\left(f_{0,k} - \frac{1}{v}\int_0^v \mathrm{d}v'\, f_{0,k}\right)\frac{|\nabla p_0|\; B_3}{BB_2}\,. \tag{58}$$

It is clear that analogous results may be obtained for higher-order moments of $\bar{f}_{1,s}$ and in particular for the kinetic energy flux.

In order to examine the structure of eq. (42) in such a case, it is furthermore convenient to write $\bar{f}_{1,s}$ as:

$$\bar{f}_{1,s} = \bar{f} = g_s + g_s^{(D)} \tag{59}$$

where $g_s^{(D)} = -\frac{|\nabla p_0|\; B_2}{\Omega_s B}\, v_{\|}\,\frac{\partial \ln N_{0,s}}{\partial p_0}\left\{1 + \eta_s\left(x_s^2 - \frac{3}{2}\right)\right\}$ is the so-called collisionless diamagnetic contribution to $\bar{f}_{1,s}$ (note that a similar definition is possible also in the case of non-symmetric configurations) and g_s satisfies the equation:

$$v_{\|}\frac{B_2}{B}|\nabla x_2|\,\frac{\partial}{\partial x_2}g_s = -\boldsymbol{v}_{D,s}\cdot\nabla f_{0,s} + \frac{e_s}{T_{0,s}}v_{\|}\; B_{\|}^{\mathrm{rot}} f_{0,s} + C_s(f_0\,|\,\bar{f}_1)\,. \tag{60}$$

We infer therefore that $\Pi_1^{(s)}$ can be written as:

$$\Pi_1^{(s)} = \Pi_{1g}^{(s)}(D) + \Pi_{1g}^{(s)} \tag{61}$$

with:

$$\Pi_{1g}^{(s)}(D) = \frac{m_s c}{e_s}\,\frac{|\nabla p_0|\; B_3}{B_2}\,\xi\,\sum_k q_{sk}\left(1 + \frac{m_s}{m_k}\right)\left\langle\frac{1}{B}\int \mathrm{d}^3v\; v_{\|}\{g_s^{(D)}\boldsymbol{V}_k - g_k^{(D)}\boldsymbol{V}_s\}\right\rangle \tag{62}$$

and analogously for $\Pi_{1g}^{(s)}$ (it sufficies to substitute $g_s \to g_s^{(D)}$ in eq. (62)). With this position, by a partial integration in x_2, an iteration can be introduced for g_s in terms of eq. (60), as for eq. (42). In particular it is immediate to prove that $\Pi_{1g}^{(s)}(D) = \Pi_{1CD}^{(s)}$ so that its previous interpretation is confirmed.

In the case of symmetric configurations the term $\Gamma_{FPS}^{(s)}$ takes a precise physical meaning. Let us consider, in fact, the limit of a weakly collisional plasma assuming $B_{\max}/B_{\min} \gg 1$ on $\Sigma(p_0)$. Then, it is obvious that $\Gamma_{FPS}^{(s)}$ survives as the leading-order contribution to the flux (since $\Gamma_{FRG}^{(s)}$ carries instead the contribution from the domain of circulating particles proportional to $B_{\min}/B_{\max}$). Similarly, it is obvious that the term $\Pi_{1PS}^{(s)}$ is non-vanishing also in the limit of a strongly collisional plasma. In analogy with ref. [14] we may therefore identify it — extending the definition to include also non-symmetric configurations — as the PFIRSCH-SCHLÜTER contribution to the flux (even if for toroidal plasmas such a definition would include an additional term proportional to q^2, with q the safety factor). Considering again as reference case toroidal axisymmetric equilibria, from eq. (48) results $\Gamma_{1RG}^{(s)} = \Pi_{1g}^{(s)}$ which is known [12] to carry — together with $\Pi_{1g}^{(s)}(D)$ — the main contribution in the limit of large aspect ratio ($R/a \gg 1$ with R and a respectively the major and minor radius of the torus). This suggests, as a consequence, that — turning back to general non-symmetric equilibria — in $\Gamma_{FRG}^{(s)}$ a qualitative distinction should be possible among its various contributions ($\Lambda_F^{(s)}$, $\Pi_{FC}^{(s)}$ and $\Pi_{FCC}^{(s)}$), e.g. whereas the first ($\Lambda_F^{(s)}$) is produced by non-conservation of $J_{||}$ on a surface $\Sigma(p_0)$, namely corresponds to a drift-motion normal to $\Sigma(p_0)$, the remaining terms are essentially produced by drift-motions on $\Sigma(p_0)$ itself.

It is interesting to examine what modifications appear in $\Pi_{FC}^{(s)}$ and $\Pi_{FCC}^{(s)}$ due to the non-conservation of $J_{||}$. A significant question, in this respect, is whether the property of strict ambipolarity is fulfilled or not by the particle fluxes.

By direct inspection, it is immediate to note that in the case of symmetric equilibria each contribution in $\Gamma_{1PS}^{(s)}$ and $\Gamma_{1RG}^{(s)}$, according to eq.s (47) and (48), fulfills such a property. In particular the 'electric' terms $\Gamma_{1E}^{(s)}$ and $\Pi_{1E}^{(s)}$ satisfy this request if the plasma is quasi-neutral at equilibrium $\left(\sum_s e_s N_{0,s} = 0\right)$, whereas for the remaining contributions it is ensured by total momentum conservation during collisions (this is obvious by inspection of eq. (62), for $\Pi_{1RG}^{(s)}$).

On the contrary, for non-symmetric equilibria, the condition of strict ambipolarity is no more automatically satisfied. More precisely it results fulfilled, under the same assumption, by $\Gamma_{1E}^{(s)}$, $\Pi_{1E}^{(s)}$ and $\Gamma_{1C}^{(s)}$, but the other terms are generally not ambipolar. This can be seen immediately in the case of $\Pi_1^{(s)}$ (eq. (25)). In fact ambipolarity would imply, from eq.s (32)—(34):

$$\sum_{s,k}\sum m_s q_{sk} \xi \left\langle \int d^3v \left(D\bar{f}_{1,s} V_{ks} - D\bar{f}_{1,k} W_{sk}\right)\right\rangle = 0 \tag{63}$$

or, with the positions $V_{ks} = v_{ks} m_s$ and $W_{sk} = w_{sk} m_s$:

$$v_{ks} = w_{ks} \quad \text{(for each species } k \text{ and } s) \tag{64}$$

which is generally not the case. A similar conclusion holds for $\Lambda_1^{(s)}$ (this requires, however, the explicit solution of the drift eq. (7)). This shows that for non-symmetric configurations ambipolarity may be ensured only by allowing for the presence of a self-consistent radial electric field. Since such a possibility has been explicitly allowed in the present ordering (see (4)) we conclude that, with an appropriate choice of the 'radial' electric field $E_r = -|\nabla p_0| \times \times \frac{\partial}{\partial p_0} \Phi_0$, ambipolarity may be ensured for the total particle flux, e.g.

$$\sum_s e_s \Gamma_1^{(s)} = 0\,. \tag{65}$$

5. Limits of Validity and a Random Walk Model

In this section we intend to examine the limits of validity of the present theory and to suggest possible generalizations as well a simple justification on the basis of the well-known random walk argument.

We note that with the SLRO we have in fact scaled (with respect to the small parameter ε_s) the frequency ratios ω_b/Ω_s, ω_{Db}/Ω_s and ν_s/Ω_s. Further possible expansions in terms of the ratio ν_s/ω_b, as corresponds to the limits $\nu_s/\omega_b \gg 1$ or $\ll 1$, must therefore be introduced consistently.

In particular, it is obvious that if $\nu_s/\omega_b \sim O(\varepsilon_s)$ or $1/O(\varepsilon_s)$ the previous derivation breaks down. Similar compatibility condition exists for ν_s/ω_{Db} and therefore we are not able to recover the result of ref. [7] which in fact deals with the assumption $\nu_s/\omega_{Db} \sim O(\varepsilon_s)$ and $\nu_s/\omega_b \sim O(\varepsilon_s)$, contrary to what explicitly requested in Sect. 2. In order to sketch how such a case can be treated with simple modifications of the present formalism, we recall the classical results reported in ref. [11] for the case of a collisionless plasma. Assuming, then, at equilibrium that $j_{||0} = 0$, integrating the drift equation along an unperturbed zeroth-order particle orbit, one obtains:

$$\frac{\partial}{\partial \beta} J_{||} \frac{\partial}{\partial p_0} f_{0,s} - \frac{\partial}{\partial p_0} J_{||} \frac{\partial}{\partial \beta} f_{0,s} = \frac{e_s}{m_s c} \oint_C ds\, C_s(f_0 \mid f_0)/v_{||}\,. \tag{66}$$

The request of local near-thermodynamic equilibrium (e.g. $f_{0,s}$ maxwellian) and non-uniform in "radial" direction $\left(\text{e.g. } \frac{\partial}{\partial p_0} f_{0,s} \neq 0\right)$, therefore, is generally not satisfied except for configurations in which $J_{||}$ is "nearly" conserved on a magnetic surface, as is the case of axisymmetric toroidal configurations with asymptotically small "ripples" superimposed on it. Assuming:

$$\frac{\partial}{\partial \beta} J_{||} \Big/ \frac{\partial}{\partial p_0} J_{||} \sim \delta \ll 1\,, \tag{67}$$

with $\varepsilon_s \ll \delta$ and expanding $f_{0,s}$ in power series of δ, to first oder in δ, eq. (60) yields:

$$\frac{\partial}{\partial \beta} J_{||} \frac{\partial}{\partial p_0} f_{0,s}^{(0)} - \frac{\partial}{\partial p_0} J_{||} \frac{\partial}{\partial \beta} f_{0,s}^{(1)} = \frac{e_s}{m_s c} \oint_C \mathrm{d}s \, C_s(f_0^{(0)} \mid f_0^{(1)})/v_{||} \tag{68}$$

where $f_{0,s}^{(0)}$ is a maxwellian distribution (here taken as constant on a magnetic surface). This eq. is equivalent to eq. (12) of ref. [7], and the present derivation shows that it is only valid for configurations for which (67) is admissible.

Instead if $\delta \sim 1$, as is the case of general non-symmetric magnetic configurations, it is obvious that a solution of eq. (60) is of the type $f_{0,s} = f_{0,s}(v, \lambda, \chi, p_0)$ and therefore — as for eq. (62) — a straightforward derivation of the fluxes shows that eq. (13) remains valid substituting for eq. (14′):

$$\Gamma_{FG}^{(s)} = \xi \langle \nabla p_0 \cdot \int \mathrm{d}^3 v \, \boldsymbol{v}_{D,s} F_s(f_{0,s} + \bar{f}_{1,s}) \rangle \tag{69}$$

where now the $O(\varepsilon_s)$-contribution due to $f_{0,s}$ is $1/O(\varepsilon_s)$ larger than that due to $\bar{f}_{1,s}$ (obviously if δ is not small with ε_s).

We want now to illustrate briefly how the previous results may be simply interpreted in terms of the stochastic random walk argument. We assume therefore that the diffusion coefficient D_s appropriate to a certain fraction f_s of the particles of species s, may be written as:

$$D_s = f_s (\Delta x_s)^2 \nu_s \,, \tag{70}$$

with Δx_s the stepsize characterizing the guiding center diffusion which occurs on the time τ_s (depending from the specific mechanism of diffusion) and ν_s the effective collision frequency. Obviously $0 < f_s \leqq 1$, whereas for the validity of the stochastic argument one must require $\Delta x_s \ll L$. In order to construct a stepsize according to the formula:

$$\Delta x_s = \text{typical velocity/typical frequency} \,, \tag{71}$$

we have to introduce a criterion of choice looking at the drift eq. (7), or eq. (66) if $\nu_s/\omega_b \sim O(\varepsilon_s)$. In the first case, since the source term of eq. (7) is proportional to the drift velocity $\boldsymbol{v}_{D,s}$ (and the term containing $E_{||}^{\mathrm{rot}}$ is of the same order of magnitude) we may identify with it the typical velocity contained in eq. (71), whereas for the frequency three choices are possible, according to the argument of Sect. 2. We obtain, therefore,

$$\Delta x_s^{(1)} \sim v_{D,s}/\nu_s \,, \qquad \Delta x_s^{(2)} \sim v_{D,s}/\omega_b \,, \qquad \Delta x_s^{(3)} \sim v_{D,s}/\omega_{Db} \tag{72′), (72″), (72‴}$$

and assume that the dominant contribution be given by trapped particles, which implies:

$$\begin{aligned} &f_s \sim \Delta; \qquad v_{||} \sim v_{\mathrm{th},s}\Delta; \qquad \omega_b \sim v_{\mathrm{th},s}\Delta/L \,, \\ &\nu_s \sim \nu_{s,0}/\Delta^2; \qquad \omega_{Db} \sim v_{\mathrm{th},s}\varepsilon_s/L; \qquad v_{D,s} \sim v_{\mathrm{th},s}\varepsilon_s \,. \end{aligned} \tag{73}$$

Here $\Delta = 1 - B_m/B_M$, where $B_m < B_M$ denote two 'a priori' arbitrary extrema of B (which may coincide, in particular with $B_{\min}$ and $B_{\max}$, the absolute extrema) on a given isobaric surface and we have still left unspecified the length L characteristic of the spatial inhomogeneities of the configuration. It follows therefore that $\Delta x_s^{(3)} (\sim L)$ is not allowed and we obtain, thus, from eq.s (72):

$$D_s^{(1)} \sim \varepsilon_s^2 \Delta^3 v_{\mathrm{th},s}^2/\nu_{s,0} \,, \qquad D_s^{(2)} \sim \Delta^{-3} r_{1,s}^2 \nu_{s,0} \,, \tag{74′), (74″}$$

which are subject to the limitations $\Delta x_s^{(i)} \ll L$ and $\nu_s/\omega_b \ll 1$, to assure the existence of trapped particles, and therefore impose, respectively, the consistency conditions:

$$\varepsilon_s \Delta^2 v_{\mathrm{th},s}/L \ll \nu_{0,s} \ll \Delta^3 v_{\mathrm{th},s}/L \,, \qquad \Delta \ll \varepsilon_s \quad \text{with} \quad \nu_{0,s} \ll \Delta^3 v_{\mathrm{th},s}/L \,. \tag{75′), (75″}$$

It is interesting to point out, furthermore, what limitations — if any — are imposed on the stochastic diffusion coefficients (74) by SLRO or more precisely — according to the argument developed at the beginning of this Section — by the request $\nu_s/\Omega_s \gg O(\varepsilon_s^2)$ necessary for the applicability of the theory developed in Sections 2—4. We infer immediately that also (74″) must fulfill the inequality on the l.h.s. of (75′), and therefore with such limitations the models (74′) and (75″) are applicable to the present transport theory. It is then possible to identify the contributions to the fluxes which carry the corresponding diffusion coefficient, either $D_s^{(1)}$ or $D_s^{(2)}$. The first, as indicated in ref. [8] for arbitrary non-symmetric equilibria and in ref. [6] for the problem of toroidal 'ripple', refers to the contribution $\Lambda_F^{(s)}$ defined by eq. (23), whereas the second appears in $\Pi_F^{(s)}$ (eq. (25)), which is — as previously recalled — the "geometrical" contribution which survives even in the case of toroidal and axisymmetric configurations and in fact one can easily recover from (74″) the usual neoclassical result [12]. Similarly with appropriate identifications of L and Δ is immediate to reproduce the results reported in [6] for the ripple problem.

Finally we mention how the weakly collisional limit $\nu_s/\omega_b \sim O(\varepsilon_s)$ corresponding to eq.s (66) and (68) may be simply included in the present stochastic model. We obtain then the constraint:

$$\nu_{s,0} \sim \varepsilon_s v_{\mathrm{th},s} \Delta^2/L \tag{76}$$

which can only be satisfied by the diffusion coefficient (74″), a result which agrees with the 'ripple' treatment of ref. [7], where an approximate solution of this problem, for asymptotically small deviations from an axisymmetric equilibrium, was considered.

6. Conclusions

A derivation of the fluxes (for particle and kinetic energy) has been presented which applies to magnetic equilibrium configurations of general type and subject to the limitations set by an appropriate asymptotic ordering in terms of the Larmor radius. In particular the properties of what has been defined as the "geometrical" contribution to the fluxes have been discussed and a partial, but still exact in terms of the unknown distribution $\bar{f}_{1,s}$, iteration has been introduced. This approach allows to express the fluxes in terms of a linear integral operator acting both on known quantities ($f_{0,s}$, $\boldsymbol{E}^{\text{rot}}$, $\boldsymbol{B}_0$) as well on $\bar{f}_{1,s}$ itself. Therefore its properties can more easily be investigated and its expression results useful for explicit determination of the fluxes in appropriate asymptotic limits. Finally we have examined the limits of validity of the present derivation and pointed out a heuristic interpretation in terms of a stochastic random walk model.

Acknowledgements

This work was performed in the framework of the author's participation in the activities of the 'Gruppo Nazionale per la Fisica Matematica del CNR' (Italy) and also as part of the resarch programme of the association agreement of Euratom and the "Stichting voor Fundamenteel Onderzoek der Materie" (FOM) with financial support from the „Nederlandse Organisatie voor Zuiver-Wetenschappelijk Onderzoek" (ZWO) and Euratom. The author is indebted to Prof. F. ENGELMANN for helpful comments and for hospitality at FOM-Instituut voor Plasmafysica and, furthermore, is obliged to Dr. J. J. LODDER for useful suggestions.

Appendix A

Here we define the coordinate systems relevant for evaluating the operators appearing in the theory.

A) The first is the orthogonal right-handed coordinate system (s, x, y) defined by HASTIE, TAYLOR and HAAS [11]. Here ds, dx and dy are the arc lengths in the directions $\boldsymbol{n}$, $\hat{e}_2$, $\hat{e}_3$ and $\hat{e}_2$, $\hat{e}_3$ need not to coincide with the principal normal and binormal. Note that since s, x, y in general are not true curvilinear coordinates the derivatives $\partial/\partial s$, $\partial/\partial x$ and $\partial/\partial y$ do not commute. For a function A independent of the azimuth θ results therefore:

$$(2\pi)^{-1} \int_0^{2\pi} d\theta\, \boldsymbol{v} \cdot \nabla A = v_{\parallel} \frac{\partial}{\partial s} A\,. \tag{A.1}$$

B) Another useful coordinate system is the generally non-orthogonal system (α, β, γ), where α and β are the EULER potentials (e.g. $\boldsymbol{B} = \nabla\alpha \times \nabla\beta$) and γ is a curvilinear coordinate on a magnetic flux line (then $\nabla\gamma \cdot \boldsymbol{n} \neq |\nabla\gamma|$ except if $\nabla \times \boldsymbol{B} = 0$ or $\boldsymbol{B} \cdot \boldsymbol{J} = 0$). If $\nabla p_0 \neq 0$ we can then identify $\alpha = p_0$. In such a case results:

$$(2\pi)^{-1} \int_0^{2\theta} d\theta\, \boldsymbol{v} \cdot \nabla A = v_{\parallel} \left(|\nabla\gamma| \frac{\partial}{\partial\gamma} + |\nabla\beta| \frac{\partial}{\partial\beta}\right) A\,. \tag{A.2}$$

If $\boldsymbol{n} \cdot \nabla f_{0,s} = 0$ we obtain then eq. (29) (note that in correspondence results $J_{\parallel} = \oint_C d\gamma\, v_{\parallel}/|\nabla\gamma|$).

C) Finally in Sect. 4 we have also considered an orthogonal right-handed system of coordinates (p_0, x_2, x_3), which is convenient in the case of equilibrium configurations with a ignorable coordinate on the isobaric surfaces. In such a case results:

$$(2\pi)^{-1} \int_0^{2\pi} d\theta\, \boldsymbol{v} \cdot \nabla A = v_{\parallel} \frac{B_2}{B} |\nabla x_2| \frac{\partial}{\partial x_2} A \tag{A.3}$$

if x_3 is the ignorable coordinate.

Appendix B

Here we report the derivation of a number of integral relationships used in Sect. 4 (e.g. eq.s (22), (28), (30) and (42)).

Let f and g be of class C^1 in the bounded domain $\Omega \cup \delta\Omega \subseteq R^3$, such that $f\nabla g$ and $g\nabla f$ are integrable on each $\omega \subseteq \Omega \cup \delta\Omega$ as well as in $\boldsymbol{v}$-space. We define:

$$g(s) = g(s_0) + \int_{C(s,s_0)} ds'\, h(s') \tag{B.1}$$

with $\oint_G ds\, h(s) \neq 0$, and $C(s, s_0)$ a piece of closed flux line C with extrema s and s_0. We intend to integrate by parts in terms of $g\boldsymbol{B} \cdot \nabla f$ the integral:

$$K_0 = \iint_{\Sigma(p_0)} d\sigma\, |\nabla p_0|^{-1} \int_0^{1/B} d\lambda\, f\boldsymbol{B} \cdot \nabla g\,. \tag{B.2}$$

One can proceed either by first using a theorem of inversion between the two integrations appearing in K_0 and then integrating by parts, or equivalently integrate by parts directly in terms of eq. (B.2). The two approaches are useful to establish a number of integral identities. We first point out the inversion theorems:

$$\iint_{\Sigma(p_0)} d\sigma\, |\nabla p_0|^{-1} \int_0^{1/B} d\lambda\, F = \int_0^{1/B_{\max}} d\lambda \iint_{\Sigma(p_0)} d\sigma\, |\nabla p_0|^{-1} F + \int_{1/B_{\max}}^{1/B_{\min}} d\lambda \sum_{i=1,k} \iint_{A_i(\lambda)} d\sigma\, |\nabla p_0|^{-1} F \tag{B.3}$$

where $B_{\max}$ and $B_{\min}$ are the absolute extrema of B on $\Sigma(p_0)$. Furthermore $A_i(\lambda)$ are the domains ($\subseteq \Sigma(p_0)$) of trapped particles, e.g. for each $i \leq k$ the regions in which $v_{\parallel}(\lambda) \geqq 0$ and constituting the ensemble of points which can be reached by a single particle with pitch-angle λ (in particular $\delta A_i(\lambda)$ is the set of points with eq. $v_{\parallel}(\lambda) = 0$ for $v \neq 0$). We integrate then by parts with the result:

$$\iint_{\Sigma(p_0)} d\sigma f\boldsymbol{B} \cdot \nabla g/|\nabla p_0| = - \iint_{\Sigma(p_0)} d\sigma\, g\boldsymbol{B} \cdot f/|\nabla p_0| + \iint_{\Sigma(p_0)} d\sigma f B[g]_{s=s_1}\, \delta(s - s_1)/|\nabla p_0| \tag{B.4}$$

and analogously for the domain $A_i(\lambda)$:

$$\iint\limits_{A_i(\lambda)} \mathrm{d}\sigma f\boldsymbol{B}\cdot\nabla g/|\nabla p_0| = \iint\limits_{A_i(\lambda)} \mathrm{d}\sigma\, g\boldsymbol{B}\cdot\nabla f/|\nabla p_0| + \iint\limits_{A_i(\lambda)} \mathrm{d}\sigma fBg\,[\delta(s - s_a(\lambda)) - \delta(s - s_b(\lambda))/|\nabla p_0| \tag{B.5}$$

where $[g]_{s=s_1} = g(s_1 + s(C)) - g(s_1)$ and $s_a(\lambda)$, $s_b(\lambda)$ are the mirror points $(v_{||}(s_a, \lambda) = v_{||}(s_t, \lambda) = 0)$.

By integrating by parts directly in terms of eq. (B.2) one obtains instead:

$$K_0 = -\xi\langle g\boldsymbol{B}\cdot\nabla g\rangle + \xi\langle fB[g]_{s=s_1}\,\delta(s - s_1)\rangle - \xi\left\langle \boldsymbol{B}\cdot\nabla\,\frac{1}{B}\,(gf)_{\lambda=1/B}\right\rangle \tag{B.6}$$

so that it is immediate to identify from (B.3) and (B.5) the third term on its r.h.s. as the contribution due to trapped particles. We obtain, namely, by making use again of the theorem of inversion (B.3):

$$\xi\left\langle \boldsymbol{B}\cdot\nabla\,\frac{1}{B}\,(gf)_{\lambda=1//B}\right\rangle = -\xi\left\langle \int\limits_{1/B_{\max}}^{1/B} \mathrm{d}\lambda\, fBg\,(\delta(s - s_a(\lambda)) - \delta(s - s_b(\lambda))\right\rangle. \tag{B.7}$$

References

1 FRIEMAN, E. A., Phys. Fluids, **13**, **490** (1970).
2 KADOMTSEV, B. B.; POGUTSE, O. P., Nucl. Fusion, **11**, 67 (1971).
3 STRINGER, T. E., Nucl. Fusion, **12**, 689 (1972).
4 CONNOR, J. W.; HASTIE, R. J., Nucl. Fusion, **13**, 221 (1973).
5 KROMMES, J. A.; RUTHERFORD, P. H., Nucl. Fusion, **14**, 695 (1973).
6 DAVIDSON, J. N., Nucl. Fusion, **16**, 731 (1976).
7 TSANG, K. T., Nucl. Fusion, **17**, 557 (1976).
8 TESSAROTTO, M., 'Ann. Mat. Pura ed Appl.' **127**, 253 (1981).
9 TESSAROTTO, M., to appear.
10 TESSAROTTO, M., submitted to 'Il Nuovo Cimento' (1981).
11 HASTIE, R. J.; TAYLOR, J. B., HAAS, F. A., Ann. Phys. , **41**, 302 (1967).
12 ROSENBLUTH, M. N.; HAZELTINE, R. D.; HINTON, F. L., Phys. Fluids, **15**, 116 (1972).
13 EDENSTRASSER, J. W., J. Plasma Phys. **24**, 299 (1980).
14 HIRSHMAN, S. P., Phys. Fluids, **19**, 155 (1976).

Eingereicht am 15. 6. 1981

Anschrift: Dr. MASSIMO TESSAROTTO, Istituto di Meccanica, Universita' degli Studi, Trieste, Italy

ZAMM 62, 533 – 538 (1982)

A. P. GUPTA / K. P. SHARMA

Axisymmetric Flexure of Circular Sandwich Plate Including Transverse Shear in Facings

Es wird eine angenäherte Formulierung für die Durchbiegung einer isotropen, kreisförmigen Sandwich-Platte unter Querbelastung vorgestellt, wobei Scherkräfte im Kern und in den Deckplatten berücksichtigt werden. Die Gleichungen werden mittels Bessel-Funktionen gelöst, und gewisse Ergebnisse für axialsymmetrische Durchbiegung berechnet, wenn der Rand eingespannt ist.

An approximate formulation is presented for the bending of an isotropic sandwich circular plate under transverse load taking into account shear forces in the core and facings. The equations are solved in terms of Bessel function and results are computed for axisymmetric bending, when the outer edge is clamped.

Представляется приближённое формулирование для прогиба изотропной круглой многослойной пластинки под поперечной нагрузкой, где принимаются во внимание усилия сдвига в ядре и на покрове. Уравнения решаются на языке функций Бесселя и вычисляются несколько результатов для осесимметричного прогиба, если обточка захватыванна.

Introduction

In the past few decades, the analysis of sandwich structure has been the topic of intensive investigation. To achieve lightness with strength, sandwich structures are being utilised in the construction of missiles and spacecrafts. A review of the work up to 1965 was done by HABIP [1] and PLANTEMMA [2]. KAO [3], has derived the system of equations for sandwich circular plate considering facings as membranes using Lagrangian multipliers, but no numerical results are presented. STICKNEY and ABDULHADI [4], have presented small deflection theory for the analysis of orthotropic circular sandwich plates by extremising the complementary strain energy. Todate solutions of circular sandwich plates are analysed by considering facings as membranes, in which transverse shear of the facings is being neglected.

In the present investigation, the axisymmetric bending of a loaded isotropic circular sandwich plate of constant thickness is considered. The transverse shear is taken into account for both core and facing. The equations of equilihrium are derived by using the principles of minimum potential energy and are solved in terms of Bessel functions. The numerical results for transverse deflection, shear forces and bending moments are computed taking outer edge clamped, both for uniform and parabolically varying load for various loadings, core thickness and facings thickness. The accuracy of the method is assessed by comparison with the known solutions, in which transverse shear of the facings is neglected.

Formulation of the problem

We consider a sandwich circular plate of thickness $2h$ and radius a. The thickness of the core and each of the two facings are taken $2h_1$ and h_2, so that $h = h_1 + h_2$. The plate is referred to cylindrical coordinates r, θ and z by taking the axis of the plate as the line $r = 0$, the middle plane as $z = 0$ and the z-axis in the downward direction. The two interfaces and bottom and top surfaces are then $z = \pm h_1$, $\pm h$, respectively. The two facings are taken of the same material, but different from that of the core. Therefore, the various quantities for the core, the lower facing and the upper facing will be distinguished by the subscripts 1, 2 and 3, respectively (Fig. 1A).

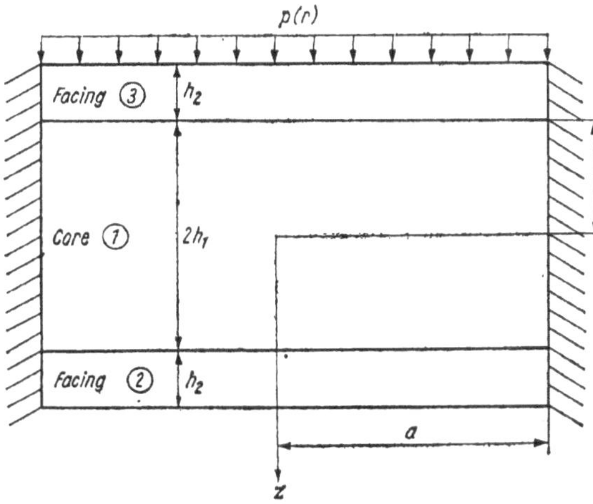

Fig. 1A. Cross-section of the plate in rz plane

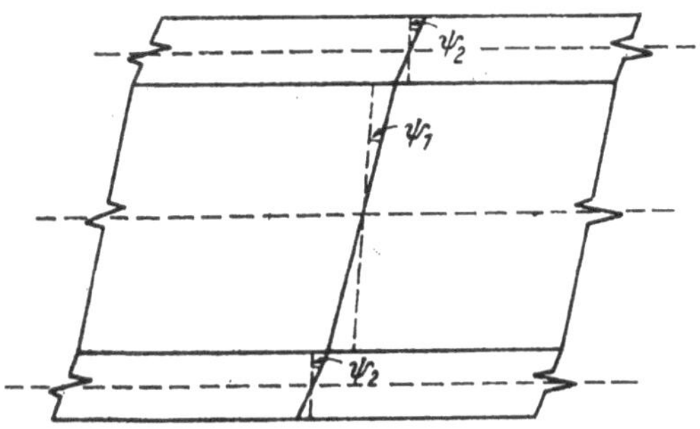

Fig. 1B. Deformation of the plate in rz plane

Since we are considering axisymmetric bending of the circular plate, the displacement in θ direction is taken to be zero. Therefore, the displacements in the core and facings are approximated as

$$u_1(r, z) = z\psi_1(r); \quad u_2(r, z) = h_1\psi_1(r) + (z - h_1)\,\psi_2(r); \tag{1}$$
$$u_3(r, z) = -h_1\psi_1(r) + (z + h_1)\,\psi_2(r); \quad w_i(r, z) = w(r); \quad v_i(r, z) = 0;$$

where $i = 1, 2$ and 3.

In the above expressions, u_i, v_i and w_i are the displacements in the directions r, θ and z, respectively, w the transverse deflection of the plate is assumed to be constant for each cross-section ($r =$ constant) and ψ_1, ψ_2 are rotations in rz-plane due to bending of the normal to the middle plane of the core and facings, respectively (Fig. 1B).

The non-zero strain components are found to be

$$\varepsilon_{r1} = z\psi_{1,r}; \quad \varepsilon_{\theta 1} = z\psi_1/r;$$
$$\varepsilon_{r2} = h_1(\psi_{1,r} - \psi_{2,r}) + z\psi_{2,r}; \quad \varepsilon_{\theta 2} = \{h_1(\psi_1 - \psi_2) + z\psi_2\}/r; \tag{2}$$
$$\varepsilon_{r3} = -h_1(\psi_{1,r} - \psi_{2,r}) + z\psi_{2,r}; \quad \varepsilon_{\theta 3} = \{-h_1(\psi_1 - \psi_2) + z\psi_2\}/r;$$
$$\varepsilon_{rz1} = \psi_1 + w_{,r}; \quad \varepsilon_{rz2} = \varepsilon_{rz3} = \psi_2 + w_{,r};$$

where a comma (,) followed by a suffix denotes differentiation with respect to that variable.

The stress-strain relations are

$$\sigma_{ri} = \lambda_i(\varepsilon_{ri} + \nu_i\varepsilon_{\theta i}); \quad \sigma_{\theta i} = \lambda_i(\varepsilon_{\theta i} + \nu_i\varepsilon_{ri}); \tag{3}$$
$$\sigma_{rzi} = \mu_i\varepsilon_{rzi}; \quad \mu_i = E_i/[2(1 + \nu_i)]; \quad \lambda_i = E_i/(1 - \nu_i^2);$$

where E_i and ν_i are the YOUNG's Modulii and POISSON's ratios, respectively.

Equations of Equilibrium

The strain energy of an isotropic sandwich plate in a polar coordinate system, neglecting the direct transverse strains ε_{zi}, may be written as

$$C = \sum_{i=1}^{3} \int_0^a \int_0^{2\pi} \int_{x_i}^{y_i} (\sigma_{ri}\varepsilon_{ri} + \sigma_{\theta i}\varepsilon_{\theta i} + \sigma_{rzi}\varepsilon_{rzi})\, r\, \mathrm{d}z\, \mathrm{d}\theta\, \mathrm{d}r; \tag{4}$$

where the limits of integration, x_i to y_i ($i = 1, 2, 3$), stand for $-h_1$ to h_1, h_1 to h, $-h$ to $-h_1$, respectively. Substituting the expressions for the strain components (2) into (4) and then performing first two integrations, the strain energy of the plate may be expressed in the form

$$C = \pi \int_0^a [rP_1\psi_{1,r} + rP_2\psi_{2,r} + P_3\psi_1 + P_4\psi_2 + rQ_rw_{,r}]\, \mathrm{d}r; \tag{5}$$

where

$$P_1 = M_{r1} - h_1N_1; \quad P_2 = M_{r2} + M_{r3} + h_1N_1; \quad P_3 = M_{\theta 1} + rQ_{r1} - h_1N_2;$$
$$P_4 = M_{\theta 2} + M_{\theta 3} + rQ_1 + h_1N_2; \quad Q_r = Q_{r1} + Q_1; \quad N_1 = N_{r2} - N_{r3}; \quad N_2 = N_{\theta 2} - N_{\theta 3};$$
$$Q_1 = Q_{r2} + Q_{r3}\,.$$

The stress-resultants are given by

$$(N_{ri}, N_{\theta i}, M_{ri}, M_{\theta i}, Q_{ri}) = \int_{x_i}^{y_i} (\sigma_{ri}, \sigma_{\theta i}, z\sigma_{ri}, z\sigma_{\theta i}, k_s\sigma_{rzi})\, \mathrm{d}z; \tag{6}$$

where limits of integration x_i, y_i are the same as for eqn. (4). Since the bending of facings and core is considered separately, we have taken the averaging shear constant $k_s = 1$ in our analysis.

The change in potential energy of the system due to applied surface load $p(r)$, is

$$U = -\int_0^a \int_0^{2\pi} p(r)\, wr\, \mathrm{d}\theta\, \mathrm{d}r\,. \tag{7}$$

The total potential energy of the system is

$$T = U + C\,. \tag{8}$$

For a system that is in equilibrium, the first variation of the total potential energy vanishes for any arbitrary set of variations of the dependent variables, w, ψ_1 and ψ_2 compatible with the prescribed boundary conditions, i.e.,

$$\delta T = \delta U + \delta C = 0\,. \tag{9}$$

Carrying out the first variation and integrating by parts those integrals containing derivatives of dependent variables, w, ψ_1 and ψ_2, we get

$$\int_0^a [(rP_{1,r} + P_1 - P_3)\,\delta\psi_1 + (rP_{2,r} + P_2 - P_4)\,\delta\psi_2 + \{rQ_{r,r} + Q_r + rp(r)\}\,\delta w]\, \mathrm{d}r +$$
$$+ [rP_1\delta\psi_1 + rP_2\delta\psi_2 + rQ_r\delta w]_0^a = 0\,. \tag{10}$$

The coefficients corresponding to variations δw, $\delta \psi_1$ and $\delta \psi_2$ must vanish, to give three equilibrium equations and vanishing of expressions already integrated with respect to r, give the edge conditions.

With the help of (3) and (6), the equations of equilibrium are given in terms of displacements by

$$2(h_1^3\lambda_1/3 + h_1^2h_2\lambda_2)(\psi_{1,rr} + \psi_{1,r}/r - \psi_1/r^2) + h_1h_2^2\lambda_2(\psi_{2,rr} + \psi_{2,r}/r - \psi_2/r^2) - 2\mu_1h_1(\psi_1 + w_{,r}) = 0; \quad (11)$$

$$h_1h_2^2\lambda_2(\psi_{1,rr} + \psi_{1,r}/r - \psi_1/r^2) + 2h_2^3\lambda_2(\psi_{2,rr} + \psi_{2,r}/r - \psi_2/r^2) - 2\mu_2h_2(\psi_2 + w_{,r}) = 0; \quad (12)$$

$$\mu_1h_1(\psi_{1,r} + \psi_1/r) + \mu_2h_2(\psi_{2,r} + \psi_2/r) + (\mu_1h_1 + \mu_2h_2)(w_{,rr} + w_{,r}/r) + p(r)/2 = 0\,. \quad (13)$$

For making above equations dimensionless, we introduce the following dimensionless variables:

$$\begin{aligned} &R = r/a; \quad H_1 = h_1/a; \quad H_2 = h_2/a; \\ &G_1 = \lambda_1/\mu_1; \quad G_2 = \lambda_2/\mu_2; \quad G_m = \mu_2/\mu_1; \\ &\mathrm{d}S_1/\mathrm{d}R = \mu_1\psi_1/P; \quad \mathrm{d}S_2/\mathrm{d}R = \mu_1\psi_2/P; \quad W = \mu_1 w/aP; \quad \bar{p}(R) = p(r)/R \end{aligned} \quad (14)$$

where P is the average load per unit area.

Eqs. (11) to (13), in dimensionless form become

$$\begin{bmatrix} a_1\nabla^2 - 2 & a_2 & -2 \\ a_3\nabla^2 & a_4\nabla^2 - 2 & -2 \\ H_1\nabla^2 & a_5\nabla^2 & a_6\nabla^2 \end{bmatrix} \begin{bmatrix} S_1 \\ S_2 \\ W \end{bmatrix} = \begin{bmatrix} 0 \\ 0 \\ -\dfrac{\bar{p}(R)}{2} \end{bmatrix}; \quad (15)$$

where $\nabla^2 \equiv \mathrm{d}^2/\mathrm{d}R^2 + \dfrac{1}{R}\,\mathrm{d}/\mathrm{d}R$

$$a_1 = 2(H_1^2G_1/3 + a_3G_m); \quad a_2 = H_2^2G_2G_m; \quad a_3 = H_1H_2G_2; \quad a_4 = 2H_2G_2/3;$$

$$a_5 = H_2G_m; \quad a_6 = H_1 + a_5\,.$$

Solving (15) for S_1, S_2 and W, we get

$$\Delta\nabla^4S_1 = (b_3\nabla^2 - 2)\,\bar{p}(R); \quad \Delta\nabla^4S_2 = (b_4\nabla^2 - 2)\,\bar{p}(R); \quad (16)\ (17)$$

$$\Delta\nabla^4W = -(b_5\nabla^4 + b_6\nabla^2 - 2)\,\bar{p}(R) \quad (18)$$

where $\Delta = b_1\nabla^2 + b_2$

$$b_1 = a_6(a_2a_3 - a_1a_4); \quad b_2 = -2(H_1b_3 - a_6b_6 + a_5b_4); \quad b_3 = a_4 - a_2; \quad b_4 = a_1 - a_3;$$

$$b_5 = -(a_1a_4 - a_2a_3)/2; \quad b_6 = a_1 + a_4\,.$$

Substituting the relations (2) and (3) in (6), we get

$$\begin{aligned} m_R = (M_{r1} + M_{r2} + M_{r3})/a^2P &= 2H_1^3G_1(S_{1,RR} + \nu_1S_{1,R}/R)/3 + \\ &+ H_2G_2G_m[H_1(H_1 + H_2)(S_{1,RR} + \nu_2S_{1,R}/R) + H_2(H_1 + 2H_2/3)(S_{2,RR} + \nu_2S_{2,R}/R); \end{aligned} \quad (19)$$

$$\begin{aligned} m_\theta = (M_{\theta1} + M_{\theta2} + M_{\theta3})/a^2P &= 2H_1^3G_1(\nu_1S_{1,RR} + S_{1,R}/R) + \\ &+ H_2G_2G_m[H_1(H_1 + H_2)(\nu_2S_{1,RR} + S_{1,R}/R) + H_2(H_1 + 2H_2/3)(\nu_2S_{2,RR} + S_{2,R}/R); \end{aligned}$$

$$q_R = (Q_{r1} + Q_{r2} + Q_{r3})/aP = 2[H_1(S_{1,R} + W_{,R}) + H_2G_m(S_{2,R} + W_{,R})]\,.$$

Solutions of Equations

The solution of eqs. (16) to (18), is composed of two parts: the complementary function and the particular integral. The latter differs for different loadings, but the former remains the same except for the arbitrary constants. The solution of eq. (16), can therefore be written as

$$S_1 = \bar{S}_1 + S_1^* \quad (20)$$

where, $\bar{S}_1$ is the complementary function and S_1^* particular integral: $\bar{S}_1$ is the solution of the equation

$$(R^2\mathrm{d}^2/\mathrm{d}R^2 + R\,\mathrm{d}/\mathrm{d}R + b_2R^2/b_1)\,\nabla^4S_1 = 0\,. \quad (21)$$

Therefore,

$$\bar{S}_1 = A_1I_0(Rn) + A_4K_0(Rn) + A_2R^2 + A_3 + A_5R^2\log R + A_6\log R \quad (22)$$

where, $n^2 = b_2/b_1$ and $A_1, \ldots, A_6$ are integration constants. I_0 and K_0 are the modified Bessel's functions of order zero. The terms containing $\log R$ and K_0 vanish for finite deflection at the centre. The complementary functions of eqs. (17) and (18), can be obtained similarly. These give

$$\bar{S}_1 = A_1I_0(Rn) + A_2R^2 + A_3; \quad \bar{S}_2 = B_1I_0(Rn) + B_2R^2 + B_3; \quad \bar{W} = C_1I_0(Rn) + C_2R^2 + C_3; \quad (23),\ (24),\ (25)$$

where, A_i, B_i, C_i ($i = 1, 2, 3$) are integration constants.

For outer edge clamped, the boundary conditions are

$$\psi_1(1) = \psi_2(1) = w(1) = 0\,. \quad (26)$$

(i) Uniform Loading: For a uniformly distributed load, we take

$$\bar{p}(R) = 1; \tag{27}$$

Eqs. (16) to (18), have then the particular integrals

$$S_1^* = S_2^* = W^* = -R^4/32b_2; \tag{28}$$

Eqs. (23) to (25), together with eq. (28) are substituted in first two eqs. (15), to obtain B_i and C_i ($i = 1, 2, 3$) in terms of A_i's and A_i's are obtained from the boundary conditions (26). These give

$$\begin{aligned} &A_1 = 2d_2/\{nI_1(n)\,(d_3 - 1)\}; \quad A_2 = d_2/(d_3 - 1) + 1/(16b_2); \\ &A_3 = C_1 I_0(n) + C_2 + 2(a_1A_2 + a_2B_2) + 1/(32b_2); \\ &B_1 = d_1A_1; \quad B_2 = d_2 + A_2; \quad B_3 = -2(b_4A_2 - b_3B_2) + A_3; \\ &C_1 = d_3A_1; \quad C_2 = -(a_1 + a_2)/(4b_2) - A_2; \quad C_3 = -2(a_1A_2 + a_2B_2) + A_3; \end{aligned} \tag{29}$$

where $I_1(n)$ is the modified Bessel's function of order one, and

$$d_1 = (n^2b_4 - 2)/(n^2b_3 - 2); \quad d_2 = (b_4 - b_3)/(4b_2);$$
$$d_3 = n^2(a_1 + a_2d_1)/2 - 1\,.$$

(ii) Parabolically Varying Load: For a load varying parabolically along the radius of the plate, we take

$$\bar{p}(R) = p_0(1 - \gamma R^2); \tag{30}$$

where, p_0 is the load intensity and γ is the rate of variation of the load. If the load varies, so that total load on the plate remains constant, then

$$\int_0^1 \int_0^{2\pi} p_0(1 - \gamma R^2)\, R\, \mathrm{d}\theta\, \mathrm{d}R = \pi P\,. \tag{31}$$

Therefore,

$$p_0/P = 2/(2 - \gamma)\,. \tag{32}$$

Eqs. (16) to (18), have the particular integrals

$$S_1^* = d_4R^4 - d_7R^6; \quad S_2^* = d_5R^4 - d_7R^6; \quad W^* = -d_6R^4 + d_7R^6; \tag{33), (34), (35}$$

where

$$d_4 = -p_0\{b_2 + 2\gamma(2b_1 + b_2b_3)\}/(32b_2^2P); \quad d_5 = -p_0\{b_2 + 2\gamma(2b_1 + b_2b_4)\}/(32b_2^2P);$$
$$d_6 = -p_0\{b_2 + 2\gamma(2b_1 + b_2b_6)\}/(32b_2^2P); \quad d_7 = -p_0\gamma/(28b_2P)\,.$$

Eqs. (23) to (25), together with eqs. (33) to (35), are solved for A_i, B_i, C_i ($i = 1, 2, 3$) as done for uniformly distributed load, to give

$$\begin{aligned} &A_1 = 2\{2(d_4 - d_5) - d_8\}/\{n(d_1 - 1)\, I_1(n)\}; \quad A_2 = -nA_1I_1(n) - 2d_4 + 3d_7; \\ &A_3 = C_1I_0(n) + C_2 + 2(a_1A_2 + a_2B_2) - d_6 + d_7; \\ &B_1 = d_1A_1; \quad B_2 = d_8 + A_2; \quad B_3 = -2(b_4A_2 - b_3B_2) + A_3; \\ &C_1 = d_3A_1; \quad C_2 = d_9 - A_2; \quad C_3 = 2(a_1A_2 + a_2B_2) - A_3; \end{aligned}$$

where

$$d_8 = -8(b_4d_4 - b_3d_5); \quad d_9 = 8(a_1d_4 + a_2d_5)\,.$$

Now, substituting the values of A_i, B_i and C_i's in the expressions for S_1, S_2 and W, we get the solutions for the two cases of uniformly distributed load and parabolically varying load.

Numerical Results and Discussions

The material for core and of the facings are taken to be cellular cellulose acetate and aluminium, for which the values of various constants are

$$G_1 = 2.20; \quad G_2 = 2.8461; \quad G_m = 1683.0;$$
$$\nu_1 = 0.0909; \quad \nu_2 = 0.2973; \quad E_1 = 5 \times 10^3\ \text{psi}; \quad E_2 = 10^7\ \text{psi}\,.$$

The transverse deflection W for various values of H_1, H_2 and γ are computed and plotted in Fig. 2 to Fig. 5. The deflection profiles for uniformly distributed load for various values of H_2, given in Fig. 2, show that the deflection decreases gradually with the gradual increase in facings thickness. The deflection profiles for uniformly distributed load for various values of H_1, given in Fig. 3, show that the deflection decreases with increase in H_1 rapidly for smaller values of H_1 and then more slowly for larger values. The deflection profiles for parabolically varying load, given in Fig. 4, show that deflection decreases as the load parameter γ decreases.

Edge moments m_R and shear forces q_R for various values of H_1, H_2 and γ are also computed. These do not change much with increase in facings or core thickness. But for parabolically varying load, the edge moments, m_R given in Table 1, for various values of load parameter γ, show a decrease with decreasing γ. Shear forces q_R vs. the

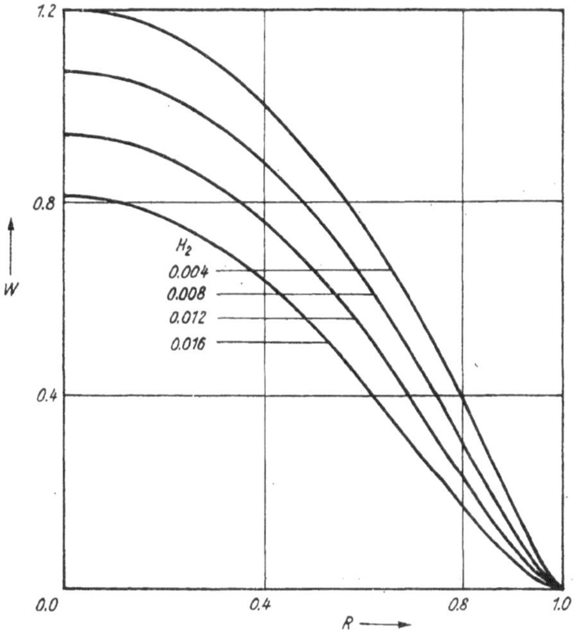

Fig. 2. Deflection profile for various facings thicknesses, $H_1 = 0.10$

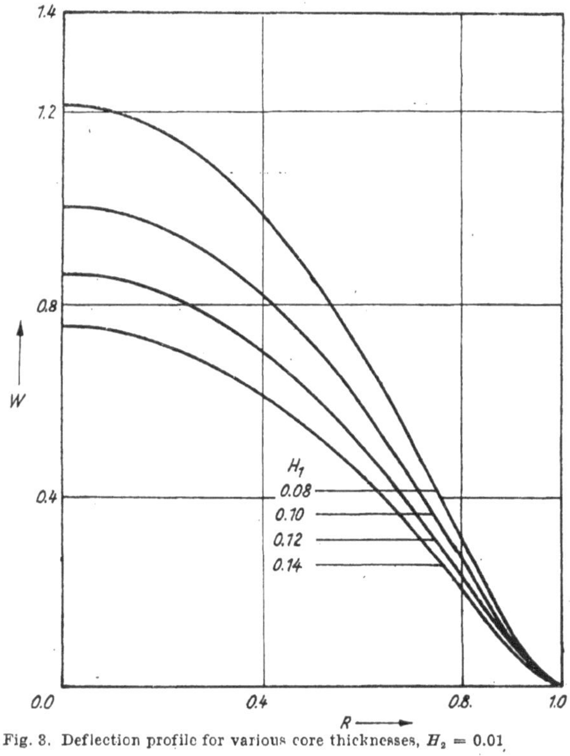

Fig. 3. Deflection profile for various core thicknesses, $H_2 = 0.01$

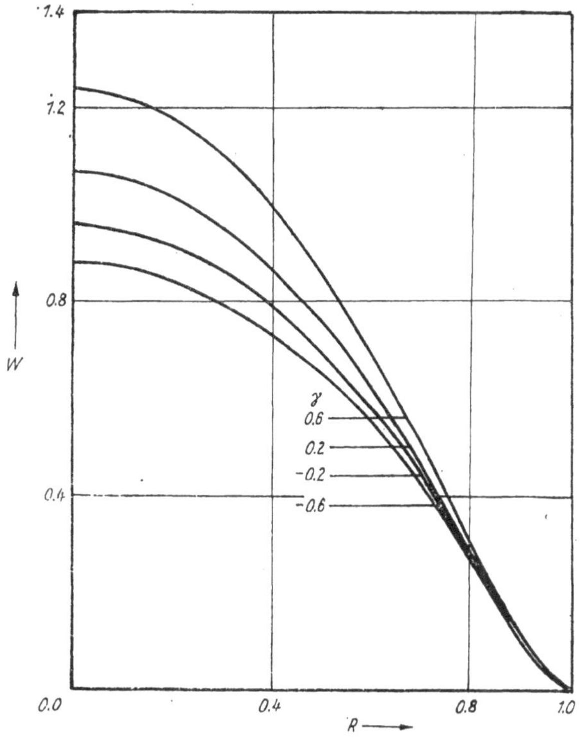

Fig. 4. Deflection profile for various values of load parameter γ, $H_1 = 0.10$, $H_2 = 0.01$

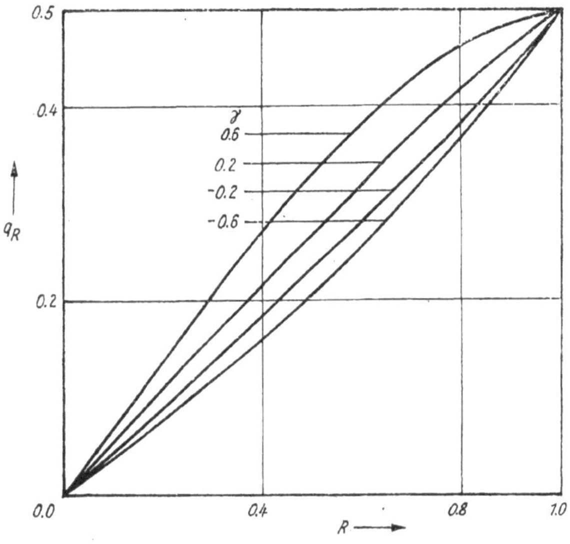

Fig. 5. Shear forces q_R vs distance from the centre for various values of load parameter γ, $H_1 = 0.10$, $H_2 = 0.01$

distance from centre for parabolically varying load for various values of γ, plotted in Fig. 5, shows that shear increases with the load concentration.

As a check, the above solution was compared with the results for circular sandwich plate, in which transverse shear of the facings is neglected. The maximum deflection at the centre, given by PLANTEMMA [2], is

$$w = \frac{Pr^4}{64D} + \frac{Pr^2}{4S} \tag{37}$$

Table 1. Edge Moments m_R for parabolically varying load, $H_1 = 0.10$, $H_2 = 0.01$

γ	Edge Moment m_R	γ	Edge Moment m_R
0.6	0.1428	−0.2	0.1212
0.2	0.1296	−0.6	0.1154

where,

$$D = \lambda_2 h_2 (2h_1 + h_2)^2/2; \qquad S = (2h_1 + h_2)^2 \, a\mu_1/(2h_1) .$$

Eq. (37), made dimensionless with the help of eq. (14), gives

$$W = [1/(16H_2G_2G_m) + H_1]/[2(2H_1 + H_2)^2] . \tag{38}$$

Using this expression to evaluate the transverse deflection of a clamped circular sandwich plate, a comparison with present analysis for various core and facings thicknesses is given in Tables 2 and 3. As a final check, we have also calculated the values for membrane theory by modifying our analysis [column 4]. The latter shows a close agreement with PLANTEMMA's value. We find a considerable difference in the transverse deflection for higher values of H_2 and smaller values of H_1. The present results show that transverse shear plays an important role, when thickness of facings is about 1/20th of core thickness or more. For such facings the error in taking the facing as membranes could exceed 5 percent.

Table 2. Transverse deflection W at the centre for uniformly distributed load, $H_1 = 0.16$

H_2	Present Theory	Membrance Theory 1 (PLANTEMMA)	Membrance Theory 2 (Present)	% Error of present theory with membrane Theory 1
0.004	0.7585	0.7776	0.7776	2.4
0.006	0.7283	0.7629	0.7629	4.5
0.008	0.6983	0.7511	0.7510	7.0
0.010	0.6675	0.7406	0.7403	9.8

Table 3. Transverse deflection W at the centre for uniformly distributed load, $H_2 = 0.004$

H_1	Present Theory	Membrane Theory 1 (PLANTEMMA)	Membrane Theory 2 (Present)	% Error of present theory with membrane Theory 1
0.08	1.4951	1.5478	1.5473	3.4
0.10	1.2025	1.2406	1.2404	3.0
0.12	1.0060	1.0351	1.0351	2.8
0.14	0.8648	0.8881	0.8880	2.6

Acknowledgements

We are grateful to Dr. C. PRASAD for his encouragement and helpful comments.

The authors would like to thank Roorkee University Regional Computer Centre for computational work. The financial support of University Grants Commission to the second author is duly acknowledged.

References

1 HABIP, L. M., A Survey of Modern Developments in the Analysis of Sandwich Structures, AMR, 18 (2), p. 93, (1965).
2 PLANTEMMA, F. J., Sandwich Construction, John Wiley and Sons, Inc., New York, (1966).
3 KAO, J. S., Bending of Circular Sandwich Plates, J. of Engg. Mech. Dn., ASCE, EM 4, pp. 165–176, (1965).
4 STICKNEY; ABDULAHDI, Flexure of Multilayer Orthotropic Circular Sandwich Plates, J. of Composite Materials, 2, 2, pp. 209–219 (1968).

Eingereicht am 15. 5. 1981

Address: Dr. K. P. SHARMA, Dr. A. P. GUPTA, Department of Mathematics, University of Roorkee, Roorkee — 247672, India

ZAMM 62, 539 – 545 (1982)

L. Lilov / M. Lorer

Dynamic Analysis of Multirigid-Body System Based on the Gauss Principle

Auf der Grundlage des Gaußschen Prinzips bringt diese Arbeit einen effizienten, rechnerorientierten Algorithmus zur Lösung der Bewegungsgleichungen eines mechanischen Mehrkörpersystems. Diesem Typ der kinematischen Kette des Systems werden keinerlei Beschränkungen auferlegt, d. h. das System darf eine beliebige Anzahl „geschlossener Kreise" enthalten. Im Kontext der Untersuchung der Dynamik von Industrie-Manipulatoren, Robotern und ähnlichen Mechanismen sind die betrachteten mechanischen Systeme gelenkige starre Körper mit Verbindungen, die Drehung und Verschiebung erlauben. Die sukzessive Berechnung basiert auf einem geeigneten Ausdruck der Gaußschen „kleinsten Restriktionen" und bedient sich der jüngst entwickelten, wirkungsvollen iterativen Programmierungsmethoden für die Minimierung einer Funktion bei Vorhandensein wie auch bei Nichtvorhandensein von Restriktionen.

This paper offers on the basis of the Gauss principle an efficient computer oriented algorithm for solving the equations of motion of multibody mechanical system. No constraints on the type of the kinematic chain of the system are imposed, i.e. the system may contain an arbitrary number of "closed loops". In the context of the investigation of the dynamics of the industrial manipulators, robots and similar mechanisms, the mechanical systems considered are linked rigid bodies with joints permitting relative rotation or translation. The successive computation is due to a suitable expression of the Gauss "leas constraint" and using of the powerful recently developed iterative programming methods for constraint and unconstrain minimization of a functional.

В работе, на основе принципа Гаусса, предложен ориентированный на применение ЭИМ эффективный алгоритм решения уравнений движения механической системы, составленной из произвольного числа тел. На тип кинематической цепи системные накладываются никакие ограничения, т.е. система может иметь произвольное число „замкнутых петель". В рамках исследования динамики промышленных манипуляторов, роботов и подобных механизмов, механические системы такого рода рассматриваются как сочлененные твердые тела с шарнирами, допускающие относительную ротацию или трансляцию. Дальнейшее эффективное вычисление возможно на основе подходящего выражения „наименьшего принуждения" Гаусса и использования недавно развитых мощных итеративных методов програмирования для минимизации функционала с ограничениями и без ограничений.

The dynamics of a multirigid body system is a subject of a number of papers such as [1]–[7] etc. There are two suitable ways for solving this basic dynamical problem, which are clearly outlined in them — knowing the external forces which act on the system they find out the generalized accelerations, velocities and coordinates. The first approach accepted in most of the papers ([1]–[5]) is based on the derivation of the nonlinear equations of motion of the mechanical system. Using these equations it's possible to carry out a great variety of investigations, connected not only with the dynamical behaviour of the system, but with its control and optimisation also.

Though, there are many problems in which the analytical investigation of the system is not necessary, but the numerical computation of the basic dynamical variables is needed, such as the determination of the motion of the system from a fixed initial state, or finding out the reactions in the joints. The second approach is more suitable to such kind of problems [6], [7]. It is based on the direct derivation of the unknown accelerations, without having to obtain an explicit form of the equations of motion. Using the Gauss principle the accelerations can be found out from the condition for minimum of a functional. The main advantage of this approach is the possibility to use effectively the mathematical programming methods, and especially the recent iterative algorithms for constraint and unconstraint minimization of quadratic functional.

Recently the rising interest towards the dynamics of the multirigid body system is connected mostly with the development of the robotics, and the necessity of precise dynamical description of the executive organs of industrial robots, the same with the biomechanical investigations of man, in particular with the mechanics of his motion.

In the contemporary technical systems most offenly used are revolute and prismatic joints. The reasons for this are obvious: these joints are easily manufactured and what is most important — the relative motion can easily be realized and controlled.

The second essential thing which has to be specially outlined is the structure of the executive organs of the industrial robots. The presence of leading bodies provides closed kinematical loops in the whole graph of the system, despite that their kinematic scheme is usually an open kinematic chain.

In this work an algorithm for a dynamical investigation of a multibody system, on the basis of the Gauss principle is proposed. No constraints on the type of the kinematic chain of the system are imposed i. e. the system may contain an arbitrary number of kinematic closed loops. The main purpose of the proposed algorithm is the investigation of the dynamics of industrial manipulators, robots and similar mechanisms. It is assumed therefore, that the joints connecting the bodies are of one degree of freedom and provide for the relative rotation or translation. The case with joints of more than one degree of freedom is treated analogously.

1. Structure of the System

The same symbolism as in paper [1] will be used. The system is supposed to have n bodies connected by $n + \hat{n}$ joints. The motion of one of the bodies is prescribed as a function of time. This body is labelled with number 0. The rest of the bodies are labeled from 1 to n, and the joints — from 1 to $n + \hat{n}$. The order in which the numbers are assigned is arbitrary, but technically is more useful to choose the so called regular labeling which is introduced in the following way:

The system can be represented by a graph Γ whose vertices and arcs symbolize the bodies and joints, respectively. It's known from the graph theory that if Γ is a cyclic graph, it can be reduced to a graph $\tilde{\Gamma}$ with a tree structure by cutting just $\hat{n}$ arcs of Γ, which generally are not uniquely choosed. In the reduced graph $\tilde{\Gamma}$, a labeling with the following properties is introduced:

1) Vertex i is incident with arc i $(i = 1, \dots, n)$,
2) If vertices i and j are incident with arc i then $i > j$ $(i, j = 1, \dots, n)$.

If the cut arcs are labeled from $n + 1$ to $n + \hat{n}$, we obtain a regularly labeled graph Γ. The bodies and joints of the system are labeled as the vertices and arcs which represent them.

If we assume, that for every two contiguous bodies a reference body for the motion of the other body is chosen, then a pair of integer functions $i^+(a)$ and $i^-(a)$, $(a = 1, \dots, n + \hat{n})$ is defined. These functions give the indices of the two bodies coupled by joint a, with $i^+(a)$ denoting the body which is chosen as a reference body. In our consideration the joints are the only revolute and prismatic, that's why there is no matter which one of the two contiguous bodies will be chosen as a reference body, so we assume that $i^+(a) < i^-(a)$.

The graph Γ can be directed by giving direction to each arc a $(a = 1, \dots, n + \hat{n})$ from vertex $i^+(a)$ to vertex $i^-(a)$. Then according to [8] the structure of Γ can be represented by the so-called incidence matrix $\underline{I}$ corresponding to the directed graph, which elements S_{ia} are defined as follows:

$$S_{ia} = \begin{cases} 1, & i = i^+(a) \\ -1, & i = i^-(a) \\ 0, & i \neq i^\pm(a) \end{cases} \qquad \begin{array}{l} i = 0, 1, \dots, n, \\ a = 1, 2, \dots, n + \hat{n}. \end{array}$$

We shall also use the submatrices $\underline{S} = (S_{ia})_{i=1,\, a=1}^{n \quad n}$, $\underline{W} = (S_{ia})_{i=1,\, a=1}^{n \quad n+\hat{n}}$, $\underline{W}_0 = (S_{0a})_{a=1}^{n+\hat{n}}$, and the matrix $\underline{W}^+$ which elements are defined by the relation:

$$S_{ia}^+ = \begin{cases} 1, & i = i^+(a) \\ 0, & i \neq i^+(a) \end{cases} \qquad \begin{array}{l} i = 1, 2, \dots, n, \\ a = 1, 2, \dots, n + \hat{n}. \end{array}$$

The inverse of matrix $\underline{S}$ is noted with $\underline{T} = (T_{ai})_{a=1,\, i=1}^{n \quad n}$. Every ordered set of indices of arcs $\langle a_1, \dots, a_k \rangle$ will be called a path between vertex i and vertex 0, if it is possible to proceed from vertex i to vertex 0 along the sequence of arcs with labels $a_1, \dots, a_k$, so that no arc is passed more than once. The path is called direct, if for all arcs of the path is true that $a_i \leqq n$. Obviously the direct path is unique.

For describing the geometric magnitudes attached to each body, and for determining its orientation in the space, it is necessary to fix vector bases with unit basis vectors $\mathbf{e}_1^{(i)}, \mathbf{e}_2^{(i)}, \mathbf{e}_3^{(i)}$, $(i = 1, \dots, n)$ rigidly attached to the bodies. Let the origin of every basis be fixed in the center of mass C_i of the corresponding body, and the basis vector $\mathbf{e}_3^{(i)}$ be parallel to the axis of joint i. The axes $\mathbf{e}_2^{(i)}$, and $\mathbf{e}_1^{(i)}$ can be chosen by different ways, but the user of the method would give preference to this one, which ensures the easiness of computation of the transformation matrices between each two contiguous bases in the reduced system. An arbitrary vector basis with origin in C_0 is attached to body 0 whose motion is a prescribed function of time, with respect to an inertial reference basis. The vector bases are denoted with the symbol $\underline{\mathbf{e}}^{(i)}$. An arbitrary vector $\boldsymbol{h}$ given in the basis $\underline{\mathbf{e}}^{(i)}$ is denoted with $\underline{h}^{(i)}$. The elements of the matrix $\underline{h}^{(0)}$ we denote with h_1, h_2, h_3.

The transformation matrix between the two-body fixed basis is defined by the equation:

$$\underline{\mathbf{e}}^{(i^-(a))} = \underline{G}_a \underline{\mathbf{e}}^{(i^+(a))}. \tag{1.1}$$

If $\langle a_1, \dots, a_k \rangle$ is the direct path for body i, and $\underline{A}_{0i}$ is the transformation matrix from $\underline{e}^{(i)}$ to $\underline{\mathbf{e}}^{(0)}$, then holds:

$$\underline{A}_{0i} = \underline{G}_{a_k} \underline{G}_{a_{k-1}} \dots \underline{G}_{a_1}, \tag{1.2}$$

An unit vector $\boldsymbol{p}_a$ and two points $D_{i^+(a)\,a}$ and $D_{i^-(a)\,a}$ are defined for each joint a $(a = 1, \dots, n + \hat{n})$, so that $\boldsymbol{p}_a$ is parallel to the joint axis, and the points $D_{i^+(a)\,a}$ and $D_{i^-(a)\,a}$ belong to the joint axis, and are rigidly attached to the bodies $i^+(a)$ and $i^-(a)$, respectively. These points are called joint points, and their radius vectors in the bases $\underline{\mathbf{e}}^{(i^+(a))}$ and $\underline{\mathbf{e}}^{(i^-(a))}$ are denoted $\mathbf{c}_{i^+(a),\,a}$ and $\mathbf{c}_{i^-(a),\,a}$ and are called joint vectors (fig. 1). We choose the joint points in the case of a revolute joint in such a way that $D_{i^+(a)\,a} = D_{i^-(a)\,a}$.

Let $\boldsymbol{R}_i$ be the radius-vector of the center of mass C_i with respect to the inertial basis, and q_a is the generalized coordinate of the relative motion of body $i^-(a)$ with respect to body $i^+(a)$, and let form the following vectors:

$$\boldsymbol{z}_a = \boldsymbol{R}_{i^-(a)} - \boldsymbol{R}_{i^+(a)} + \mathbf{c}_{i^-(a),\,a} - \mathbf{c}_{i^+(a),\,a}. \tag{1.3}$$

Equality (1.3) shows that after the prescribed choice of the joint points is made, the following conditions hold:

$$\boldsymbol{z}_a = \boldsymbol{0} \quad \text{if} \quad \mu_a = 0, \tag{1.4}$$

$$\boldsymbol{z}_a \times \boldsymbol{p} = \boldsymbol{0} \quad \text{if} \quad \mu_a = 1 \tag{1.5}$$

where

$$\mu_a = \begin{cases} 1 & \text{if joint } a \text{ is prismatic}, \\ 0 & \text{if joint } a \text{ is revolute} \end{cases} \qquad a = 1, \dots, n + \hat{n}.$$

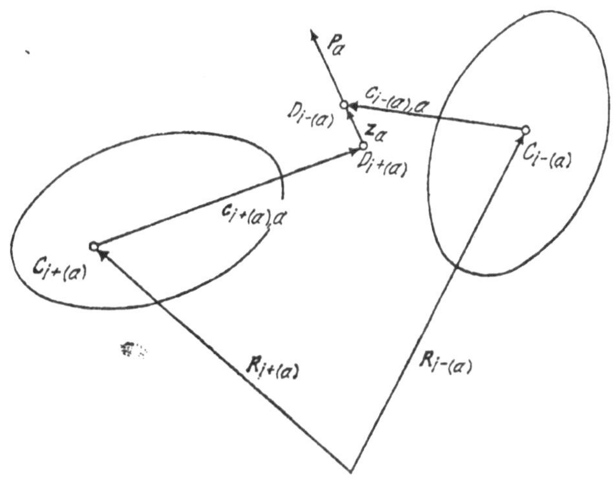

Fig. 1

Eqs. (1.4) and (1.5) can be combined in one equality:

$$z_a = \mu_a p_a q_a\,, \qquad a = 1, \dots, n + \hat{n}\,. \tag{1.6}$$

Let $\boldsymbol{\Omega}_a$ be the vector of the relative angular velocity of body $i^-(a)$ relative to $i^+(a)$. Obviously:

$$\boldsymbol{\Omega}_a = \boldsymbol{\omega}_{i^-(a)} - \boldsymbol{\omega}_{i^+(a)} = (1 - \mu_a)\, \boldsymbol{p}_a \dot{q}_a \tag{1.7}$$

where $\boldsymbol{\omega}_i$ is the absolute angular velocity of body i, and the symbol $(\dot{\ })$ is used for the absolute time derivative. Using the matrix $\underline{T}$ it's possible to solve eqs. (1.4), (1.5) for $a = 1, \dots, n$, in regard to $\boldsymbol{R}_i$ and $\boldsymbol{\omega}_i$, respectively, as it is done in [2], and to obtain:

$$\boldsymbol{R}_i = -\sum_{a=1}^{n} T_{ai}(\boldsymbol{c}_{i^+(a),a} - \boldsymbol{c}_{i^-(a),a} + \mu_a \boldsymbol{p}_a q_a) + \boldsymbol{R}_0\,, \tag{1.8}$$

$$\boldsymbol{\omega}_i = -\sum_{a=1}^{n} T_{ai}(1 - \mu_a)\, \boldsymbol{p}_a \dot{q}_a + \boldsymbol{\omega}_0\,. \tag{1.9}$$

After the differentiation of eq. (1.8), a formula for the linear velocity of the center of mass C_i is found:

$$\dot{\boldsymbol{R}}_i = -\sum_{a=1}^{n} T_{ai}[(\boldsymbol{\omega}_{i^+(a)} \times \boldsymbol{c}_{i^+(a),a}) - (\boldsymbol{\omega}_{i^-(a)} \times \boldsymbol{c}_{i^-(a),a})] + \mu_a \boldsymbol{p}_a \dot{q}_a + \mu_a(\boldsymbol{\omega}_{i^+(a)} \times \boldsymbol{p}_a)\, q_a + \dot{\boldsymbol{R}}_0\,. \tag{1.10}$$

2. The Least Constraint Principle of Gauss

In any dynamical system let $\boldsymbol{r}_i$ is the radius-vector of a typical particle m_i at time t, and $\boldsymbol{F}_i$ is the external force which acts on it. Consider function β which is called "constraint"

$$\beta = \frac{1}{2} \sum_i \frac{1}{m_i} (\boldsymbol{F}_i - m_i \ddot{\boldsymbol{r}}_i)\, (\boldsymbol{F}_i - m_i \ddot{\boldsymbol{r}}_i)\,. \tag{2.1}$$

The summation is extended over all particles of the system. Using this function GAUSS formulated the following principle:

"At any moment of time the actual motion of a mechanical system, subordinated to ideal constraints, differs from all other kinematicly possible motions by the property, that for it β obtains minimal value".

For a system of n rigid bodies eq. (2.1) takes the form:

$$\beta = \frac{1}{2} \sum_{i=1}^{n} \int_{m_i} \frac{1}{\mathrm{d}m} (\mathrm{d}\boldsymbol{F} - \ddot{\boldsymbol{r}}\, \mathrm{d}m)\, (\mathrm{d}F - \ddot{\boldsymbol{r}}\, \mathrm{d}m)\,. \tag{2.2}$$

Let $\tilde{\boldsymbol{F}}_i$ be the resultant force on body i, and $\tilde{\boldsymbol{M}}_i$ the resultant torque about i-th body center of mass. Let them be represented in the form:

$$\tilde{\boldsymbol{F}}_i = \boldsymbol{F}_i + \boldsymbol{F}_i^*\,, \qquad \tilde{\boldsymbol{M}}_i = \boldsymbol{M}_i + \boldsymbol{M}_i^*$$

where $\boldsymbol{F}_i$ is the resultant external force, and $\boldsymbol{M}_i$ the resultant external torque about C_i, and $\boldsymbol{F}_i^*$ is the resultant internal force, and $\boldsymbol{M}_i^*$ the resultant internal torque about C_i.

The inertial characteristics of a typical body i are given by its mass m_i and inertia tensor $\mathbb{I}_i$ about its center of mass. If $\ddot{\boldsymbol{R}}_i$ is the absolute linear acceleration of C_i, and $\dot{\boldsymbol{\omega}}_i$ is the absolute angular acceleration of body i, than the following form for the "constraint" β can be obtained:

$$\beta = \tfrac{1}{2} \sum_{i=1}^{n} [\ddot{\boldsymbol{R}}_i m_i \ddot{\boldsymbol{R}}_i + \dot{\boldsymbol{\omega}}_i \mathbb{I}_i \dot{\boldsymbol{\omega}}_i + 2\dot{\boldsymbol{\omega}}_i(\boldsymbol{\omega}_i \times \mathbb{I}_i \boldsymbol{\omega}_i) - 2\ddot{\boldsymbol{R}}_i(\boldsymbol{F}_i^* + \boldsymbol{F}_i) - 2\dot{\boldsymbol{\omega}}_i(\boldsymbol{M}_i^* + \boldsymbol{M}_i)] . \tag{2.3}$$

This form of β is derived from eq. (2.2) after the substitution:

$$\ddot{\boldsymbol{r}} = \ddot{\boldsymbol{R}}_i + \ddot{\boldsymbol{\varrho}}$$

in which $\boldsymbol{\varrho}$ denotes the radius-vector of the mass element dm in the basis $\underline{e}^{i^+(a)}$ (fig. 2).

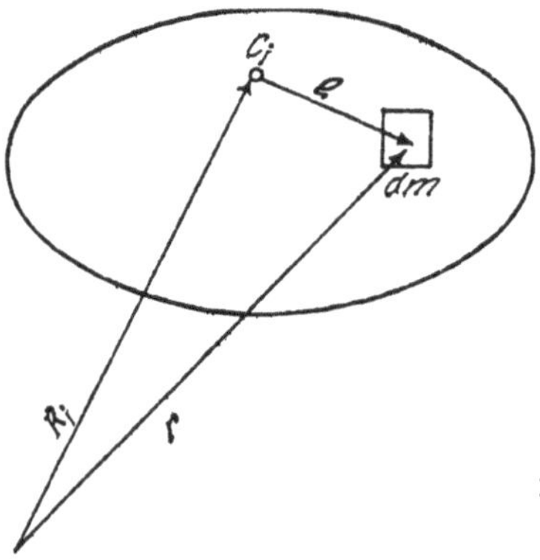

Fig. 2

In accordance with the GAUSS principle the minimum of β is to be found out when a fixed configuration and fixed velocities in this configuration are supposed, i.e. the unknown variables of the minimisation problem are $\ddot{\boldsymbol{R}}_i$ and $\dot{\boldsymbol{\omega}}_i$, $i = 1, \dots, n$. In case of a multirigid body system the minimum is conditional, because of the constraints imposed on the motion of the bodies. These constraints are reduced to the conditions, that the relative motion of any two contiguous bodies is simply rotation or translation. Using the already introduced notations, and the symbol (°) for the local derivative of an arbitrary vector in the basis $\boldsymbol{e}^{i^+(a)}$ we can write down the constraints in case of a relative rotation in the joint a

$$\mathring{\boldsymbol{z}}_a = \boldsymbol{0} , \qquad \boldsymbol{\Omega}_a \times \boldsymbol{p}_a = \boldsymbol{0} \tag{2.5), (2.6}$$

and in case of a relative translation in the joint a

$$\boldsymbol{\Omega}_a = \boldsymbol{0} , \qquad \mathring{\boldsymbol{z}}_a \times \boldsymbol{p}_a = \boldsymbol{0} . \tag{2.7), (2.8}$$

Equalities (2.5—2.8) are equivalent to the following:

$$\boldsymbol{\Omega}_a \times \boldsymbol{p}_a = \boldsymbol{0} , \qquad \boldsymbol{z}_a = \boldsymbol{0} , \qquad \boldsymbol{\Omega}_a = \boldsymbol{0} , \qquad \boldsymbol{z}_a \times \boldsymbol{p}_a = \boldsymbol{0} . \tag{2.5*), (2.6*), (2.7*), (2.8*}$$

It is obvious that $\boldsymbol{z}_a = \boldsymbol{0}$ yields $\mathring{\boldsymbol{z}}_a = \boldsymbol{0}$, and $\boldsymbol{z}_a \times \boldsymbol{p}_a = \boldsymbol{0}$ yields $\mathring{\boldsymbol{z}}_a \times \boldsymbol{p}_a + \boldsymbol{z}_a \times \mathring{\boldsymbol{p}}_a = \boldsymbol{0}$. But $\mathring{\boldsymbol{p}}_a = 0$, hence (2.8) is true. On the other side integrating (2.6) and (2.8) and having in mind the already made choice of the joint points, one can find out that the constants of the integration are zero vectors, hence (2.6*) and (2.8*) are true.

Eq. (2.5* —2.8*) can be written in more compact way:

$$\mu_a \boldsymbol{z}_a \times \boldsymbol{p}_a + (1 - \mu_a)\, \boldsymbol{z}_a = \boldsymbol{0} , \qquad (1 - \mu_a)\, \boldsymbol{\Omega}_a \times \boldsymbol{p}_a + \mu_a \boldsymbol{\Omega}_a = \boldsymbol{0} . \tag{2.9}$$

In order to obtain the constraints upon the unknown accelerations $\ddot{\boldsymbol{R}}_i$ and $\dot{\boldsymbol{\omega}}_i$ ($i = 1, 2, \dots, n$) eq. (2.9) is differentiated

$$\begin{gathered} \mu_a \ddot{\boldsymbol{z}}_a \times \boldsymbol{p}_a + (1 - \mu_a)\, \ddot{\boldsymbol{z}}_a + \mu_a \boldsymbol{z}_a \times \ddot{\boldsymbol{p}}_a + 2\mu_a \dot{\boldsymbol{z}}_a \times \dot{\boldsymbol{p}}_a = \boldsymbol{0} , \\ (1 - \mu_a)\, \dot{\boldsymbol{\Omega}}_a \times \boldsymbol{p}_a - \mu_a \dot{\boldsymbol{\Omega}}_a - (1 - \mu_a)\, \boldsymbol{\Omega}_a \times \dot{\boldsymbol{p}}_a = \boldsymbol{0} \end{gathered} \tag{2.10}$$

and eqs. (1.3), (1.7) are used represented in the following way:

$$-\boldsymbol{z}_a = \sum_{i=0}^{n} S_{ia}(\boldsymbol{R}_i + \boldsymbol{c}_{ia}) , \qquad -\boldsymbol{\Omega}_a = \sum_{i=0}^{n} S_{ia} \boldsymbol{\omega}_i , \qquad a = 1, \dots, n + \hat{n} . \tag{2.11}$$

The vectors $\boldsymbol{c}_{ia}$ have no meaning in the case when $S_{ia} = 0$, i.e. when $i \neq i^{\pm}(a)$. For defineteness we shall suppose that if $S_{ia} = 0$, then $c_{ia} = 0$. When we substitute (2.11) in (2.10) and having in mind that

$$\dot{\boldsymbol{p}}_a = \boldsymbol{\omega}_{i^+(a)} \times \boldsymbol{p}_a = \sum_{i=0}^{n} S_{ia} \boldsymbol{\omega}_i \times \boldsymbol{p}_a , \qquad \dot{\boldsymbol{c}}_{ia} = \boldsymbol{\omega}_i \times \boldsymbol{c}_{ia}$$

we find the following final form of the constraints:

$$\sum_{i=1}^{n} S_{ia} \mu_a \boldsymbol{p}_a \times \ddot{\boldsymbol{R}}_i - \sum_{i=1}^{n} S_{ia}(1 - \mu_a)\, \ddot{\boldsymbol{R}}_i + \sum_{i=1}^{n} S_{ia} \mu_a \boldsymbol{p}_a \times \dot{\boldsymbol{\omega}}_i - \sum_{i=1}^{n} S_{ia}(1 - \mu_a)\, \dot{\boldsymbol{\omega}}_i - \sum_{i=1}^{n} S_{ia}^{+} \boldsymbol{z}_a \times \\ \times (\mu_a \boldsymbol{p}_a \times \dot{\boldsymbol{\omega}}_i) + \boldsymbol{u}_a = \boldsymbol{0} \tag{2.12}$$

where

$$u_a = S_{0a}\mu_a p_a \times \ddot{R}_0 - S_{0a}(1-\mu_a)\ddot{R}_0 + S_{0a}\mu_a p_a \times \dot{\omega}_0 - S_{0a}(1-\mu_a)\dot{\omega}_0 -$$
$$- S_{0a}^{+} z_a \times (\mu_a p_a \times \dot{\omega}_0) - \sum_{i=0}^{n} S_{ia}\mu_a[\omega_i \times (\omega_i \times c_{ia})] \times p_a - \sum_{i=0}^{n} S_{ia}(1-\mu_a)[\omega_i \times (\omega_i \times c_{ia})] +$$
$$+ \mu_a z_a \times [\omega_{i^+(a)} \times (\omega_{i^+(a)} \times p_a)] + 2\mu_a \dot{z}_a \times (\omega_{i^+(a)} \times p_a), \tag{2.13}$$

$$\sum_{i=1}^{n} S_{ia}(1-\mu_a)\, p_a \times \dot{\omega}_i - \sum_{i=1}^{n} S_{ia}\mu_a \dot{\omega}_i + v_a = 0 \tag{2.14}$$

where

$$v_a = S_{0a}(1-\mu_a)\, p_a \times \dot{\omega}_0 - S_{0a}\mu_a\dot{\omega}_0 + \sum_{i=1} S_{ia}(1-\mu_a)(\omega_{i^+(a)} \times p_a) \times \omega_i . \tag{2.15}$$

We write down eqs. (2.12—2.15) in the basis $\underline{e}^{(0)}$, using the transformation matrices $\underline{A}_{0i}$ $(i = 1, \dots, n)$ and the transformation matrix from the inertial basis to $\underline{e}^{(0)}$, which is supposed to be known at every instant of time t. After we have written down the constraints for $a = 1, \dots, n + \hat{n}$ in the joints, we can determine the accelerations $\underline{\ddot{R}}_i^{(0)}$ and $\underline{\dot{\omega}}_i^{(0)}$ $(i = 1, \dots, n)$ at the moment t, from the minimum conditions for β, subject to the constraints (2.12), (2.14) for each joint and for fixed $\underline{\dot{R}}_i^{(0)}$, $\underline{R}_i^{(0)}$, $\underline{\omega}_i^{(0)}$ $(i = 1, \dots, n)$ at the considered instant of time.

Integrating $\underline{\ddot{R}}_i^{(0)}$ twice and $\underline{\dot{\omega}}_i^{(0)}$ once, we obtain $\underline{\dot{R}}_i^{(0)}$, $\underline{R}_i^{(0)}$ and $\underline{\omega}_i^{(0)}$ respectively at the next instant of time $t + \Delta t$. The angular orientation of the body i in relation with the zero one, we can calculate in the following way: by $\boldsymbol{\Omega}_{0i}$ we denote the vector of the angular velocity of body i relative to the body zero. It's obvious that:

$$\dot{e}_k^{(i)} = \Omega_{0i} \times e_k^{(i)} . \tag{2.16}$$

Then by $A_{0i}^{[k]}$ we denote the k-th column of the matrix $\underline{A}_{0i}$, and by $\underline{\tilde{\Omega}}_{0i}$ the matrix:

$$\underline{\tilde{\Omega}}_{0i} = \begin{bmatrix} 0 & -\Omega_{0i}^3 & \Omega_{0i}^2 \\ \Omega_{0i}^3 & 0 & -\Omega_{0i}^1 \\ -\Omega_{0i}^2 & \Omega_{0i}^1 & 0 \end{bmatrix}. \tag{2.17}$$

Then (2.16) and (2.17) yield:

$$\dot{A}_{0i}^{[j]} = \underline{\tilde{\Omega}}_{0i}^{(0)} A_{0i}^{[j]} . \tag{2.18}$$

The solution of the system of differential equations (2.18) for $j = 1, 2, 3$ yields the matrix $\underline{A}_{0i}$ at the moment $t + \Delta t$.

In addition to the transformation matrices, it is necessary to obtain the values of the generalized velocities $\dot{q}_a$ and coordinates q_a. For this purpose we use the eqs. (1.6), (1.7) and the equation

$$\dot{z}_a = \mu_a p_a \dot{q}_a + \mu_a q_a(\omega_{i^+(a)} \times p_a)$$

projected in the basis $\underline{e}^{(0)}$. Knowing $\underline{\omega}_i^{(0)}$, $\underline{R}_i^{(0)}$, $\underline{\dot{R}}_i^{(0)}$, $\underline{p}_a^{(0)}$, $\underline{c}_{i^+(a),a}^{(0)}$ and $\underline{c}_{i^-(a),a}^{(0)}$, we can calculate $\dot{q}_a$ in the case of rotation using any of the equalities

$$\dot{q}_a = \dot{q}_a^i = (\omega_{i^-(a)}^i - \omega_{i^+(a)}^i)/p_a^i , \qquad i = 1, 2, 3; \qquad p_a^i \neq 0 . \tag{2.19}$$

Integrating $\dot{q}_a$ we obtain the generalized coordinate q_a. In the case of translation analogously, q_a and $\dot{q}_a$ can be calculated, using any of the equalities.

$$q_a = q_a^i = z_a^i/p_a^i , \qquad i = 1, 2, 3; \qquad p_a^i \neq 0 , \tag{2.20}$$
$$\dot{q}_a = \dot{q}_a^i = [\dot{z}_a^i - (\underline{\tilde{\omega}}_{i^+(a)}^{(0)} \underline{p}_a^{(0)})^i q_a]/p_a^i , \qquad i = 1, 2, 3; \qquad p_a^i \neq 0 \tag{2.21}$$

where

$$\underline{\tilde{\omega}}_i^{(0)} = \begin{bmatrix} 0 & -\omega_i^3 & \omega_i^2 \\ \omega_i^3 & 0 & -\omega_i^1 \\ -\omega_i^2 & \omega_i^1 & 0 \end{bmatrix}.$$

It remains to obtain the forces F_i^* and the torques M_i^* $(i = 1, \dots, n)$ in order to form the „constraint" β at the moment $t + \Delta t$. That is done in the following way.

Let X_a is the resultant vector with respect to $C_{i^-(a)}$, of all the internal forces doing virtual work, which act in joint a on body $i^+(a)$. Then on body $i^-(a)$ acts force $-X_a$. Using the matrix $\underline{W}$ it can be shown, that the resultant force F_i of all the internal forces in the joints connected with the i-th body is [1]:

$$F_i^* = \sum_{a=1}^{n+\hat{n}} W_{ia} X_a .$$

Let Y_a be the sum of all the torques about the point $D_{i^-(a)a}$, corresponding to all the internal forces acting in the joint a on body $i^+(a)$. Then on body $i^-(a)$ acts a torque $-Y_a$. Hence the torques of the internal forces in the joint a about the centres of mass $C_{i^+(a)}$ and $C_{i^-(a)}$ are

for body $i^+(a)$:

$$Y_a - (c_{i^+(a)} + z_a) \times X_a ,$$

for body $i^-(a)$:

$$-\boldsymbol{Y}_a - (\boldsymbol{c}_{i^-(a)} \times \boldsymbol{X}_a)\,.$$

Let:

$$\boldsymbol{c}_{ia} = \begin{cases} \boldsymbol{c}_{i^+(a),\,a}\,, & i = i^+(a) \\ \boldsymbol{c}_{i^-(a),\,a}\,, & i = i^-(a)\,, \\ 0 \quad , & i \neq i^\pm(a) \end{cases} \qquad a = 1, \dots, n + \hat{n}; \qquad i = 0, \dots, n\,,$$

$$\boldsymbol{c}^*_{ia} \doteq \begin{cases} \boldsymbol{z}_a\,, & i = i^+(a)\,, \\ 0\,, & i \neq i^+(a)\,, \end{cases} \qquad a = 1, \dots, n + \hat{n}; \qquad i = 0, \dots, n\,.$$

One can prove that the torque about C_i of all the internal forces in all joints connected with body i is:

$$\boldsymbol{M}^*_i = \sum_{a=1}^{n+\hat{n}} W_{ia}[\boldsymbol{Y}_a + (\boldsymbol{c}_{ia} + \boldsymbol{c}^*_{ia}) \times \boldsymbol{X}_a]\,.$$

In order to finish the description of the algorithm it is necessary to make clear its first step, which significantly differs from the others. The cause for this is that usually the initial configuration of the system and initial velocities $\boldsymbol{\omega}_i$, $\dot{\boldsymbol{R}}_i$ $(i = 1, \dots, n)$ are not known. Instead of them, we usually know the values of the generalized coordinates and velocities. In order to obtain the initial values for $\boldsymbol{\omega}_i$, $\boldsymbol{R}_i$, $\dot{\boldsymbol{R}}_i$, $\underline{A}_{0i}$ $(i = 1, \dots, n)$ which are necessary for the algorithm, we use eqs. (1.9), (1.8), (1.10), (1.2) respectively. These equations are no more used, after the first step of the algorithm is completed.

3. Solving the Minimisation Problem

In Section 2 it is shown, that using the Gauss principle, the problem of finding out the accelerations in the multirigid body system, is brought to the standard problem of finding out the minimum of a quadratic functional. In order to obtain β in the standard mode we make the following notations:

$$\underline{x}^T = (\ddot{R}_1^{(0)T}, \ddot{R}_2^{(0)T}, \dots, \ddot{R}_n^{(0)T}, \dot{\omega}_1^{(0)T}, \dot{\omega}_2^{(0)T}, \dots, \dot{\omega}_n^{(0)T})\,,$$

$$\underline{Q} = \operatorname{diag}\,(m_1\underline{E}_3, m_2\underline{E}_3, \dots, m_n\underline{E}_3, I_1^{(0)}, I_2^{(0)}, \dots, I_n^{(0)})\,,$$

$$\underline{B} = (\underline{\widetilde{F}}_1^{(0)}, \dots, \underline{\widetilde{F}}_n^{(0)}, \underline{\widetilde{M}}_1^{(0)} + \underline{\theta}_1^{(0)}, \dots, \underline{\widetilde{M}}_n^{(0)} + \underline{\theta}_n^{(0)})$$

where $I_j^{(0)}$ is the inertia tensor of body j in the basis $\underline{e}^{(0)}$, E_3 is the (3×3) identity matrix and $\underline{\theta}_k^{(0)}$ $(k = 1, \dots, n)$ are given with the equalities:

$$\underline{\theta}_k^{(0)} = \underline{\widetilde{\omega}}_k^{(0)}\, I_k^{(0)} \underline{\omega}_k^{(0)}\,, \qquad k = 1, \dots, n\,.$$

Equations (2.12), (2.14) are linear with respect to the accelerations, hence they can be written down in the way:

$$\underline{A}\underline{x} = \underline{b}\,.$$

In order to obtain the matrix $\underline{A}$, we define the following matrices:

$$\underline{\widetilde{p}}_a^{(0)} = \begin{bmatrix} 0 & -p_a^3 & p_a^2 \\ p_a^3 & 0 & -p_a^1 \\ -p_a^2 & p_a^1 & 0 \end{bmatrix}, \qquad \underline{\widetilde{z}}_a^{(0)} = \begin{bmatrix} 0 & -z_a^3 & z_a^2 \\ z_a^3 & 0 & -z_a^1 \\ -z_a^2 & z_a^1 & 0 \end{bmatrix},$$

$$\underline{\widetilde{P}}^{(0)} = \operatorname{diag}\,(\underline{\widetilde{p}}_1^{(0)}, \underline{\widetilde{p}}_2^{(0)}, \dots, \underline{\widetilde{p}}_{n+\hat{n}}^{(0)})\,, \qquad \underline{\widetilde{Z}}^{(0)} = \operatorname{diag}\,(\underline{\widetilde{z}}_1^{(0)}, \underline{\widetilde{z}}_2^{(0)}, \dots, \underline{\widetilde{z}}_{n+\hat{n}}^{(0)})\,,$$

$$\underline{A}_1 = \operatorname{diag}\mu\,\underline{\widetilde{P}}^{(0)}\underline{W}^T - (\underline{E}_{n+\hat{n}} - \operatorname{diag}\mu)\,\underline{W}^T\,,$$

$$\underline{A}_2 = \operatorname{diag}\mu\,\underline{\widetilde{P}}^{(0)}\underline{W}^T - (\underline{E}_{n+\hat{n}} - \operatorname{diag}\mu)\,\underline{W}^T - \operatorname{diag}\mu\,\underline{\widetilde{Z}}^{(0)}\underline{W}^{+T}\,,$$

$$\underline{A}_3 = (\underline{E}_{n+\hat{n}} - \operatorname{diag}\mu)\,\underline{\widetilde{P}}^{(0)}\underline{W}^T - \operatorname{diag}\mu\,\underline{W}^T\,.$$

The matrix $\underline{A}$ then receives the following bloc structure:

$$\underline{A} = \left[\begin{array}{c|c} \underline{A}_1 & \underline{A}_2 \\ \hline 0 & \underline{A}_3 \end{array}\right].$$

The matrix $\underline{b}$ is used in the sense:

$$\underline{b}^T = \left(\underline{u}_1^{(0)T}, \underline{v}_1^{(0)T}, \underline{u}_2^{(0)T}, \underline{v}_2^{(0)T}, \dots, \underline{v}_{n+\hat{n}}^{(0)T}, \underline{v}_{n+\hat{n}}^{(0)T}\right)$$

where $\underline{u}_a^{(0)}$ and $\underline{v}_a^{(0)}$ $(a = 1, \dots, n + \hat{n})$ are obtained from (2.13) and (2.15).

Using the introduced notations we obtain from (23) the following mathematical programming problem:

$$\beta = f(\underline{x}) = \underline{x}^T\underline{Q}\underline{x} + \underline{B}\underline{x} \to \min \tag{3.1}$$

subject to the constraints:

$$\underline{A}\underline{x} = \underline{b}\,. \tag{3.2}$$

In order to simplify the notations, let $m = 6(n + \hat{n})$ and $s = 6n$. Then $\underline{Q}$ has dimensions $(s \times s)$, and $\underline{A}(m \times s)$.

The considered problem has one important peculiarity — the matrix $\underline{A}$ is singular and $m \geqq s$. This complicates the computational solving of the problem. In order to surmount these difficulties we decided to apply the conjugate gradient projection method, which seems to be mostly convinient for the case — see e.g. [10].

As it is known from the linear algebra, the matrix $A^+(s \times m)$ is called pseudoinverse of the matrix $A(m \times s)$, if the following conditions hold:

$$\underline{A}^+\underline{A}\underline{A}^+ = \underline{A}^+\,, \qquad \underline{A}\underline{A}^+\underline{A} = \underline{A}\,, \qquad (\underline{A}^+\underline{A})^T = \underline{A}^+\underline{A}\,, \qquad (\underline{A}\underline{A}^+)^T = \underline{A}\underline{A}^+\,. \tag{3.3}$$

The operator $P\colon R^s \to R^s$ is called projector if:

$$\underline{P}.\underline{P} = \underline{P}\,, \qquad \underline{P}^* = \underline{P} \tag{3.4}$$

where P^* is the conjugate operator. Using (3.3) one can directly prove that the operator

$$\underline{P} = \underline{A}^+\underline{A} \tag{3.5}$$

satisfies eq. (3.4), hence it is projector. It is easy to see, that the general solution of eq. (3.2) is given with:

$$\underline{x} = \underline{x}_0 + (\underline{E}_s - \underline{P})\,\underline{y} \tag{3.6}$$

where $\underline{y}$ is an arbitrary s-dimensional vector, $\underline{P}$ is the projector given with the eq. (3.5) and $\underline{x}_0$ is given with

$$\underline{x}_0 = \underline{A}^+\underline{b}\,. \tag{3.7}$$

With the help of the replacement (3.6) the conditional minimization problem, is brought to the folloing unconditional minimization problem [10]

$$f(\underline{x}_0 + (\underline{E}_s - \underline{P})\,\underline{y}) \to \min\,.$$

4. On the Computational Realization of the Method

Obviously, the practical effectiveness of the algorithm depends on its numerical realization. In the sequel, we give some suggestions, which have been used in our computational experiments at the ES-1040 computer.

1) The proper work of the algorithm depends essentially on the accuracy of the integration scheme. Therefore the numerical integration should be rather precise. We have used the ADAMS-MULTON scheme with variable step, combined with some corrector-predictor formulas.

2) Various geometric and kinematic variables are successively transformed in the basis $\underline{e}^{(0)}$. Thus for reducing the error, one should periodically orthonormalize the transformation matrices $\underline{A}_{0i}$ obtained by eq. (2.18).

3) For obtaining the pseudoinverse matrix $\underline{A}^+$, the singular expansion of the matrix $\underline{A}$ can be used. An idea for such an algorithm is presented in [11], where a standard subprogram SVD is used. The subprogram SVD however, is more general than needed for computing $\underline{A}^+$. For example, using SVD one can find the eigenvectors and eigenvalues of the matrices $\underline{A}^+\underline{A}$ and $\underline{A}\underline{A}^+$. Therefore, for increasing the speed of computation, one can suggests the direct calculation of $\underline{A}^+$, by means of GREVIL method [12].

4) Each of equalities (2.5*) and (2.8*) defines three scalar kinematic constraints, two of which are lineary independent. For eliminating the dependent constraint we propose the following procedure:

1. Order by value $|p_a^1|$, $|p_a^2|$, $|p_a^3|$ for each $a = 1, 2, \ldots, n + \hat{n}$ and let $|p_a^i| \leqq |p_a^j| \leqq |p_a^k|$ $i, j, k \in \{1, 2, 3\}$.
2. Check:

   if $\mu_a = 0$ then $\lambda_a = 6a - 3 + k$, if $\mu_a = 1$ then $\lambda_a = 6a - 6 + k$.
3. Reduce the matrices $\underline{A}$ and $\underline{b}$ by eliminating the rows with indices $\lambda_1, \lambda_2, \ldots, \lambda_{n+\hat{n}}$.

   In such a way the dimensions of the matrix $\underline{A}$ becomes $5(n + \hat{n}) \times 6n$.

References

1 LILOV, L.; WITTENBURG, J., Bewegungsgleichungen für Systeme starrer Körper mit Gelenken beliebiger Eigenschaften, ZAMM **57**, 1977.
2 WITTENBURG, J., Dynamics of Systems of Rigid Bodies, Teubner, Stuttgart 1977.
3 Лилов, Л., Върху динамиката на едноконтурни системи, Теоретична и приложна механика — под печат.
4 GUPTA, V., Dynamic Analysis of Multirigid Body System, ASME Journal of Engineering for Industry, Vol. **96** 1974.
5 UICKER, J. J., Jr., Dynamic Behavior of Spatial Linkages, ASME Paper **96** MECH. 1.
6 Верещагин, А. Ф., Принцип Гауса для моделирования на ЦВМ динамики роботов-манипуляторов, ДАНСССР т. 220 №1 1975.
7 Попов, Е.; Верещагин, А.; Зенкевич, С., Манипулационные роботы" Наука 1978.
8 HARARY, F., Graph Theory, Adison-Wesley Pub. Company 1969.
9 WHITTAKER, E., Analytical Dynamics of Particles and Rigid Bodies, Cambridge University Press 1965.
10 Пшеничный, Б.; Данилин, Ю., Численные методы в екстремальных задачах, Наука 1965.
11 FORSYTHE, J.; MALCOLM, M.; MOLER, Cl., Computer Methods for Mathematical Computations, Prentice Hall 1979.
12 Гантмахер, Ф., Теория матриц, Наука 1966.

Eingereicht am 21. 8. 1981

Anschrift: Doz. Dr. L. LILOV, Mat. M. LORER, Institute of Mechanics and Biomechanics, Bulgarian Academy of Sciences, 1090 Sofia, P. O. Box 373, Bulgaria.

ZAMM 62, 547–560 (1982)

G. BOESE

Einschlüsse und Trennung der Nullstellen von Exponentialtrinomen mit konstanten Koeffizienten

Es werden Exponentialtrinome $A \exp(az) + B \exp(bz) + C \exp(cz)$ *in der komplexen Variablen* z *mit komplexen konstanten Koeffizienten und Exponentenpunkten betrachtet. Alle ihre Nullstellen werden mehrfach eingeschlossen. Sind die Exponentenpunkte nicht kollinear, so können fast alle Nullstellen durch Kreisscheiben getrennt werden. Kenntnis der Nullstellen ist für die Lösungs- und Stabilitätstheorie zu Differenzengleichungen bedeutsam.*

Exponential trinomials $A \exp(az) + B \exp(bz) + C \exp(cz)$ *in the complex variable* z *with complex constant coefficients and exponential points are considered. All their zeros are included by several inclusion regions. For non co-linear exponential points almost all zeros are separated by disks. The distribution of the zeros is essential in the theory of difference equations, especially in the related stability theory.*

Рассматриваются экспоненциальные трехчлены вида $A \exp(az) + B \exp(bz) + C \exp(cz)$ в комплексной переменной z с постоннными комплексными коэффициентами и экспоненциальными точками. Включаются все нули с помощью различных областей вклюнения. Если экспоненциальные точки неколлинеарны, потом почти все нули разделимы кругами. Распределение нулей существенно в теории решения и устойчивости разностных уравнений.

1. Aufgabenstellung und Motivation

Ziel dieser Arbeit ist es, sämtliche Nullstellen von Exponentialtrinomen

$$T(z) := A_1 \exp(a_1 z) + A_2 \exp(a_2 z) + A_3 \exp(a_3 z)\,, \tag{1.1}$$

in der komplexen Variablen $z \in \mathbb{C}$ mit konstanten, nicht verschwindenden, $A_1 A_2 A_3 \neq 0$, komplexen Koeffizienten $A_k \in \mathbb{C}$ und mit voneinander verschiedenen, $(a_1 - a_2)(a_1 - a_3)(a_2 - a_3) \neq 0$, komplexen Exponentenpunkten $a_k \in \mathbb{C}$ durch gut zu überblickende und aufwandsarm zu berechnende Einschlußmengen realistisch zu lokalisieren.

Es gibt mancherlei Veranlassung, die Nullstellenverteilung von T mit $n \in \mathbb{N}$ Summanden anstatt dreier Summanden wie in (1.1) und mit polynomialen Koeffizienten $A_k \in \mathbb{C}[z]$ zu betrachten. Derartige Exponentialpolynome sind die charakteristischen Funktionen zu über $\mathbb{R}$ erklärten Differenzen-Differentialgleichungen mit konstanten Koeffizienten und konstanten Argumentverschiebungen. In der Lösungs- und Stabilitätstheorie zur genannten Gleichungsklasse benötigt man die Nullstellen der zugehörigen charakteristischen Gleichung bzw. gewisse Informationen über die Verteilung der Nullstellen. Unter Verzicht auf größere Genauigkeit kann man (in nicht ganz trivialer Weise) den Fall polynomialer Koeffizienten auf den konstanter zurückführen.

Bei Diskretisierungen von partiellen Differentialoperatoren mit konstanten Koeffizienten in $\mathbb{R}^2$ gelangt man zu reinen Differenzengleichungen mit in $\mathbb{C}$ liegenden Argumentverschiebungen. Man denke an bekannte Differenzensterne. Im Zuge der Analyse numerischer Verfahren, die Differenzengleichungen benutzen, sind die Nullstellen der mit den Diskretisierungen verbundenen Exponentialpolynomen mit konstanten Koeffizienten zu ermitteln.

Nichtidentisch verschwindende Monome haben keine Nullstellen. Binome verteilen ihre Nullstellen äquidistant längs Geraden in $\mathbb{C}$. Der erste nichttriviale Fall tritt bei den Trinomen T auf.

Gewöhnlicherweise wird die Lage der Nullstellen von Exponentialpolynomen mit der von Exponentialbinomen verglichen. Vergleicht man mit Trinomen, so gelangt man zu gesteigerter Genauigkeit und zu einem größeren Gültigkeitsbereich für den Vergleich. Das Vorgehen ist allerdings nur dann von praktischem Interesse, wenn man die Lage der Nullstellen der Trinome gut kennt. Dem Verfasser ist nicht bekannt, daß die Nullstellen der Trinome T in der reichhaltigen, sich über ein knappes Jahrhundert erstreckenden Literatur behandelt worden wären.

Man kann T durch ein Laplace-Integral

$$T(z) := (2\pi i)^{-1} \int_C \exp(zt)\, P_2(t)\, \big((t - a_1)(t - a_2)(t - a_3)\big)^{-1}\, dt\,, \tag{1.2}$$

wobei P_2 ein Polynom zweiten Grades ist, mit passendem Integrationsweg C darstellen. Aus (1.2) wird einsichtig, daß bei der asymptotischen Auswertung von komplizierten Laplace-Integralen durch Abspaltung des wesentlichen Trinoms eine engere Approximation an das auszuwertende Integral und ein erweiterter Gleichmäßigkeitsbereich bezüglich vorhandener Parameter erreicht wird, als dies bei einer monomialen oder binomialen Approximation möglich wäre. Praktikabel ist das Vorgehen wieder nur dann, wenn Trinome T wie das Monom $\exp(z)$ und das Binom $\cos(z)$ zu den elementaren Transzendenten gezählt werden dürfen. Mit der Verteilung der Nullstellen ist in der Klasse der Exponentialpolynome die ganze Wertverteilung bekannt.

Der Fall gewöhnlicher Trinome in der Variablen $\exp(z)$ und der Fall von Dirichlet-Trinomen, wobei $a_k := \ln(n_k)$, $n_k \in \mathbb{N}_0$, zu nehmen ist, sind in (1.1) enthalten.

Es werden mehrere Einschlüsse angegeben. Unter Punkt 2 wurden polygonal berandete Einschlüsse aufgeführt. Sie sind so einfach gehalten, daß sie mit Zirkel und Lineal (und einer Rechenhilfe) in $\mathbb{C}$ eingezeichnet werden können. Für große $|z|$ läuft die Einschlußmenge in Parallelstreifen aus. Die konstante Streifenbreite wird in Punkt 3 zu exponentiell fallender Streifenbreite herabgedrückt. Ein zweiter Einschluß, der in einer Klasse von Fällen zum ersten fast orthogonal verläuft, ist in Punkt 4 dargestellt. In Punkt 5 werden Kreisscheibeneinschlüsse fast aller

Nullstellen für eine weite Trinomklasse abgeleitet. Eine Diskussion unter Punkt 6 schließt die Arbeit ab. Bilder illustrieren gewonnene Resultate.

Die Exponentenpunkte fassen wir in der Menge $Q := \{a_1, a_2, a_3\}$ zusammen. Die konvexe Hülle conv (Q) von Q hat zwei oder drei Eckpunkte. Umlaufen wir das Randpolygon ∂ conv (Q) von conv (Q) im Gegenuhrzeigersinn, so erscheinen die Ecken von conv (Q) in einer bestimmten zyklischen Reihenfolge. Die Eckpunktmenge von conv (Q), in dieser Reihenfolge aufgeschrieben, heiße der *Eckpunktzyklus* $V(Q) := (a, b, c)$ oder $V(Q) := (a, b)$, falls conv (Q) eine zweiseitige Strecke ist. Der rechte Nachbar des letzten Eckpunktes in V ist wieder der erste Eckpunkt. Die lexikographische Ordnung der Eckpunkte gebe immer die Ordnung in V wider. Das zu T gehörige Indikatordiagramm ist conv $(\overline{Q})$, wobei $\overline{Q}$ die an der reellen Achse gespiegelte Menge Q ist. Die Koeffizienten werden künftig mit dem zugehörigen Exponentenpunkt indiziert.

Es bezeichnet stets $i := \sqrt{-1}$, und $H(t) := 1$ für $t \geqq 0$ sowie $H(t) := 0$ für $t < 0$ ist der über $\mathbb{R}$ definierte *Einheitssprung*.

2. Ein polygonal berandeter Einschluß mit konstanter Streifenbreite

Für das Folgende ist es angebracht, die Koeffizienten durch

$$|A_a| + |A_b| + |A_c| = 1 \tag{2.1}$$

zu normieren.

Im Konflikt zwischen Genauigkeit und Einfachheit der Einschlußmengen entscheiden wir uns anfangs für die Einfachheit, um einen Überblick zu bekommen. Wir lassen vorerst nur polygonal berandete Einschlußmengen zu.

Satz 2.1:

a) *Besitzt* conv (Q) *nur zwei Eckpunkte a und c, während b innerer Punkt ist, so liegen alle Nullstellen von T im Parallelstreifensystem*

$$\left.\begin{aligned}
&S := \{z \in \mathbb{C} \mid z = w/(c-a), \operatorname{Re}(w) \in U, \operatorname{Im}(w) \in \mathbb{R}\}\,,\\
&U := [U_1, U_2] \cup [U_3, U_4]\,, \quad U_1 := \min(L_a, L_a/l)\,, \quad U_2 := -\max(0, L_b/l)\,,\\
&U_3 := \max\left(0, L_b/(1-l)\right), \quad U_4 := \max\left(-L_c, -L_c/(1-l)\right),\\
&L_a := \ln\left(|A_a|/(1-|A_a|)\right), \quad a \in Q\,, \quad l := (b-a)/(c-a) > 0\,.
\end{aligned}\right\} \tag{2.2}$$

Für $L_b > 0$ zerfällt S in zwei Parallelstreifen.

b) *Hat* conv (Q) *drei Eckpunkte, so liegen alle Nullstellen von T im Parallelstreifensystem*

$$\left.\begin{aligned}
&S := D \cup S_{a,b} \cup S_{b,c} \cup S_{c,a}\,,\\
&D := \operatorname{conv}(\{z_a, z_b, z_c\})\,, \quad S_{a,b} := \{z \in \mathbb{C} \mid z = (1-u)\, z_b + u z_c + iv, 0 \leqq u \leqq 1, v \in \mathbb{R}_+\}\,, \quad a \in Q\,,\\
&z_a := L_a s_a\,, \quad a \in Q\,, \quad s_a := 1/\left(i(c-b) \operatorname{Im}\left((b-a+c-a)/(c-b)\right)\right),\\
&L_a := \ln\left(|A_a|/(1-|A_a|)\right).
\end{aligned}\right\} \tag{2.3}$$

Die Streifenbreite von $S_{a,b}$ ist

$$d_{a,b} := |L_a + L_b|/|b-a|\,. \tag{2.4}$$

Die Randhalbgeraden von $S_{a,b}$ verlaufen senkrecht zur Seite conv $(\{a, \bar{b}\})$ *des Indikatordiagramms von T und in Richtung zur Außennormale dieser Seite ins Unendliche.*

In Bild 1 ist der von Satz 2.1 gewährte Nullstelleneinschluß für das Trinom

$$T(z) := \left(\exp(az) + \exp(-az) + \exp(bz)\right)/3\,, \qquad a := (1+i)/\sqrt{2}\,, \qquad b := (-1+i)/\sqrt{2}\,, \tag{2.5}$$

gezeichnet. Neben dem Rand von S ist der Rand des Indikatordiagramms dargestellt. Die Mittelpunkte der Nullkreise sind die Ecken des Indikatordiagramms.

Beweis: Wir wählen $a \in Q$ beliebig aus und schreiben die Nullstellengleichung in bezug auf diese Auswahl mit $Q_a := Q \setminus \{a\}$ in der Form

$$A_a \exp(az) = -\sum_{a \in Q_a}' A_b \exp(bz)\,, \qquad A_a = -\sum_{a \in Q_a}' A_b \exp\left((b-a)\, z\right). \tag{2.6}$$

Jede Nullstelle z von T genügt (2.6) und somit erst recht

$$\begin{aligned}
&|A_a| \leqq (1-|A_a|) \exp\left(\max\{\operatorname{Re}((b-a)\, z) : b \in Q_a\}\right),\\
&L_a := \ln\left(|A_a|/(1-|A_a|)\right) \leqq \max\{\operatorname{Re}((b-a)\, z) : b \in Q_a\}\,.
\end{aligned} \tag{2.7}$$

In (2.7) wurde die Koeffizientennormierung (2.1) benutzt. Im Rest des Beweises wird das System der für alle $a \in Q$ betrachteten Ungleichungen der letzten Zeile von (2.7) nach z aufgelöst und eine geeignete Darstellung der Lösungsmenge angegeben.

Wir beginnen mit dem Fall, daß conv (Q) nur die Eckpunkte a und c besitzt, während b innerer Punkt ist. Mit dem Ansatz $z := w/(c-a)$, $w := u + iv$, $u, v \in \mathbb{R}$, bekommt das genannte System die Gestalt

$$\left.\begin{aligned}
&L_a \leqq \max(lu, u)\,, \qquad L_b \leqq \max\left(-lu, (1-l)\, u\right),\\
&L_c \leqq \max\left(-u, -(1-l)\, u\right), \qquad l := (b-a)/(c-a)\,.
\end{aligned}\right\} \tag{2.8}$$

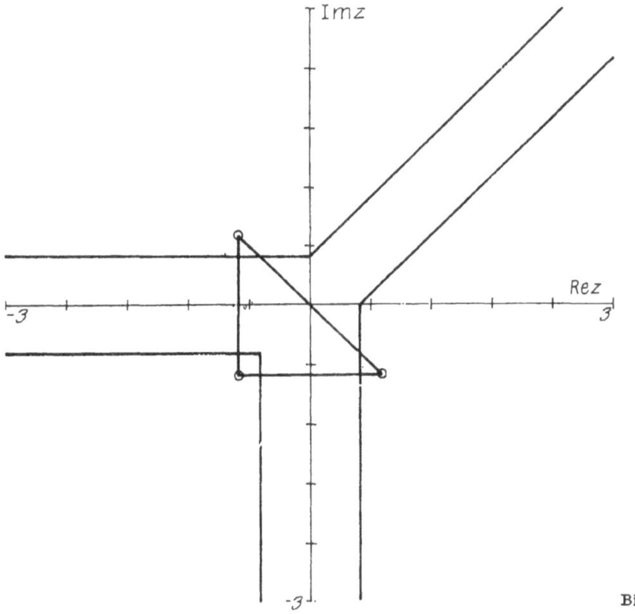

Bild 1. Der Einschluß von Satz 2.1 zum Trinom T nach (2.5)

Man findet die Lösungsmenge U von (2.8) elementar wie im Satz angegeben.

Hat conv (Q) drei Eckpunkte, so setzen wir

$$u := \operatorname{Re}\big((b-a)\,z\big)\,,\qquad v := \operatorname{Re}\big((c-a)\,z\big) \tag{2.9}$$

an und gelangen zum System

$$L_a \leqq \max(u, v)\,,\qquad L_b \leqq \max(-u, v-u)\,,\qquad L_c \leqq \max(-v, u-v)\,. \tag{2.10}$$

Die Menge S' der Punkte $(u, v) \in \mathbb{R}^2$, die (2.10) lösen, läßt, die Darstellung

$$\left.\begin{aligned} &S' := D' \cup S'_{a,b} \cup S'_{b,c} \cup S'_{c,a}\,,\qquad D' := \operatorname{conv}(\{s'_a, s'_b, s'_c\})\,,\\ &S'_{b,c} := \{(u, v) \in \mathbb{R}^2 \mid (u, v) \in \operatorname{conv}(\{s'_b, s'_c\}) + (t, t), t \in \mathbb{R}_+\}\,,\\ &S'_{c,a} := \{(u, v) \in \mathbb{R}^2 \mid (u, v) \in \operatorname{conv}(\{s'_c, s'_a\}) - (t, 0), t \in \mathbb{R}_+\}\,,\\ &S'_{a,b} := \{(u, v) \in \mathbb{R}^2 \mid (u, v) \in \operatorname{conv}(\{s'_a, s'_b\}) - (0, t), t \in \mathbb{R}_+\}\,,\\ &s'_a := (L_a, L_a)\,,\qquad s'_b := (-L_b, 0)\,,\qquad s := (0, -L_c) \end{aligned}\right\} \tag{2.11}$$

zu. Die Parallelstreifen $S'_{a,b}$ bis $S'_{c,a}$ entstehen also aus den Seiten des Randdreiecks von D' durch Translation längs der in (2.11) angegebenen Richtungen. Aus der Normierung (2.1) folgt, daß für jedes Paar (a, b) in V aufeinanderfolgender Ecken $L_a + L_b < 0$ erfüllt ist. Das bedeutet, daß die Streifenbreiten der $S'_{a,b}$ bis $S'_{c,a}$ positiv sind. Mit der affinen Abbildung (2.9) wird die Menge S' in die z-Ebene rückübertragen und dort als S bezeichnet. Entsprechend auch für Untermengen von S'. Aus den Eigenschaften affiner Abbildungen folgt, daß S ebenfalls ein zusammenhängendes Parallelstreifensystem ist. Es wird im Satz ohne Rückgriff auf das Urbild S' beschrieben. Die drei Eckpunkte z_a von S sind Lösungen von

$$L_a = \operatorname{Re}\big((b-a)\,z\big) = \operatorname{Re}\big((c-a)\,z\big). \tag{2.12}$$

Um sicherzustellen, daß $S_{a,b}$ aus der Translation der Seite conv $(\{z_a, z_b\})$ in Richtung der Außennormale zur Seite conv $(\{\bar{a}, \bar{b}\})$ des Indikatordiagramms entsteht, wurde gebraucht, daß a, b, c im Zyklus V aufeinander folgen. Q.e.d.

Satz 2.1 erlaubt die folgende Konstruktionsvorschrift für den *Rand der Einschließungsmenge*:

1. Man zeichne den Rand des Indikatordiagramms.
2. Man zeichne die Eckpunkte z_a, z_b, z_c der Einschlußmenge.
3. Man zeichne von den Punkten z_a und z_b ausgehende Halbstrahlen in der Richtung der Außennormale zur Seite conv $(\{\bar{a}, \bar{b}\})$ des Indikatordiagramms. Verfahre ebenso mit den verbleibenden beiden Eckpunktpaaren.

Für den Fall *kollinearer Exponentenpunkte* verbessern wir Satz 2.1. Im jetzt betrachteten Fall kann man das Trinom auf die Gestalt (2.13) bringen.

Satz 2.2: *Die Nullstellen des Trinoms*

$$T(z) := 2\operatorname{ch}(z/2) + P\exp(pz/2)\,, \tag{2.13}$$

mit $-1 < p < 1$ *und* $P \in \mathbb{C}$ *liegen im Vertikalstreifensystem*

$$S := \{z \in \mathbb{C} \mid z := |x|\,\sigma + iy,\ |x| \in X(\sigma),\ y \in \mathbb{R},\ \sigma \in \{-1, 1\}\},$$

$$X(\sigma) := \begin{cases} [0, A(\sigma)], & 0 < q < 1/2, \quad 0 \leqq Q \leqq Q_0, \\ [0, B(\sigma)] \cup [C(\sigma), A(\sigma)], & 0 < q < 1/2, \quad Q_0 \leqq Q \leqq 2, \\ [C(\sigma), A(\sigma)], & 0 < q \leqq 1/2, \quad 2 \leqq Q, \\ [D(\sigma), A(\sigma)], & 1/2 \leqq q < 1; \end{cases}$$

$$A(\sigma) := \ln\left(Q + \min(1, Q^{1-1/q})\right)/q, \qquad B(\sigma) := V(2-Q)/(2-Q_0),$$

$$C(\sigma) := V + \ln\left((Q/Q_0)/(1 - q + q(Q_0/Q)^{1/q})\right)/q,$$

$$D(\sigma) := \begin{cases} 0, & 0 \leqq Q \leqq 2, \\ \ln\left(Q/(1 + (2/Q)^{1/q})\right)/q, & 2 \leqq Q; \end{cases}$$

$$V := \ln(1/q - 1), \qquad Q_0 := \left((1-q)/q\right)^q/(1-q), \qquad q := (1 - p\sigma)/2, \qquad Q := |P|.$$

In Abhängigkeit von $|P|$ *verteilen sich die Nullstellen von* T *also auf einen oder zwei Vertikalstreifen.*

Beweis: Zweckmäßigerweise beziehen wir die z-Ebene auf kartesische Koordinaten $z := x + iy$, $x, y \in \mathbb{R}$. Nach Betragsbildung in (2.13) und Quadrieren bekommen wir für jede Nullstelle z von T mit $c := \cos(y)$

$$\exp(|x|) + 2c + \exp(-|x|) = |P|^2 \exp(px). \tag{2.14}$$

Aus $-1 \leqq c \leqq 1$ folgt mit $Q := |P|$ sowie $q := (1 - p\sigma)/2$ und $\sigma := \operatorname{sgn}(x)$ und $t := |x|$

$$f_1(t) := \exp(qt)\left(1 - \exp(-t)\right) \leqq Q \leqq \exp(qt)\left(1 + \exp(-t)\right) =: f_2(t). \tag{2.15}$$

Aus $-1 < p < 1$ ergibt sich $0 < q < 1$.

Wir verschaffen uns zunächst einen strukturellen Überblick über die Lösungsmenge T von (2.15). Die Funktion f_1 wächst für die zugelassenen q monoton, $f_1(\mathbb{R}_+) = \mathbb{R}_+$. Also gibt es für alle $Q \in \mathbb{R}_+$ genau ein Lösungsintervall $[0, t_1]$. Die Funktion f_2 hat unterschiedliches Monotonieverhalten. Für $0 < q < 1/2$ besitzt f_2 ein einziges Minimum an der Stelle $t_0 := \ln(1/q - 1) =: V$ in der Höhe $Q_0 := \left((1-q)/q\right)^q/(1-q)$.

Wegen $f_2(0) = 2$ existiert das links von t_0 liegende Lösungsintervall nur für $Q_0 \leqq Q \leqq 2$, das rechte für alle $Q \geqq Q_0$. Für $q \geqq 1/2$ ist $f_2 \geqq 2$ und monoton wachsend. Es gibt somit für $Q \geqq 2$ ein einziges Lösungsintervall. Für die Lösungsmenge T von (2.15) gilt

$$T := \begin{cases} [0, t_1], & 0 < q < 1/2, \quad 0 \leqq Q < Q_0, \\ [0, t_{2,1}] \cup [t_{2,2}, t_1], & 0 < q < 1/2, \quad Q_0 \leqq Q \leqq 2, \\ [t_{2,2}, t_1], & 0 < q < 1/2, \quad 2 \leqq Q < \infty, \\ [t_2, t_1], & 1/2 \leqq q < 1. \end{cases} \tag{2.16}$$

Die t_k in (2.16) sind die Lösungen von $f_k(t) = Q$, $k = 1, 2$. Mehrere Lösungen sind durch einen zweiten Index unterschieden. Die Lösungen t sind so abzuschätzen, daß die Intervalle von T nicht verkleinert werden.

Wir beginnen mit $f_1 = Q$ und schreiben diese Gleichung in passender Form und schätzen ab

$$\begin{aligned} &t = \ln\left(Q + \exp(-t + qt)\right)/q, \qquad t \geqq \max(0, \ln(Q)/q), \\ &t_1 \leqq \ln\left(Q + \min(1, Q^{1-1/q})\right)/q := A(\sigma). \end{aligned} \tag{2.17}$$

Für $Q \geqq 2$ und $q \geqq 1/2$ ist die einzige Lösung t_2 von $f_2 = Q$ minorant abzuschätzen. Es ist

$$t = \ln\left(Q/(1 + \exp(-t))\right)/q \geqq \ln(Q/2)/q, \qquad t_2 \geqq \ln\left(Q/(1 + (2/Q)^{1/q})\right)/q.$$

Somit ist für alle Q

$$t_2 \geqq D(\sigma) := \begin{cases} 0, & 0 \leqq Q \leqq 2, \\ \ln\left(Q/(1 + (2/Q)^{1/q})\right)/q, & 2 \leqq Q. \end{cases} \tag{2.18}$$

In ihrem Existenzbereich ist die Wurzel $t_{2,1}$ majorant abzuschätzen. Für $0 < q < 1/2$ ist $f_2'' > 0$ in ganz $\mathbb{R}_+$ und $f_2' \leqq 0$ in $[0, t_0]$. Die Sehne, die die Funktionswerte von f_2 an den Stellen $t = 0$ und $t = t_0$ miteinander verbindet, liegt somit oberhalb des Graphen von f_2,

$$Q = f_2(t) \leqq f_2(0) - (t/t_0)\left(f_2(0) - f_2(t_0)\right) = 2 - (t/t_0)(2 - Q_0), \qquad t_{2,1} \leqq t_0(2-Q)/(2-Q_0) =: B(\sigma). \tag{2.19}$$

Schließlich ist $t_{2,2}$ minorant abzuschätzen. Für $t \geqq t_0$ gilt

$$\left.\begin{aligned} &t = \ln\left(Q/(1 + \exp(-t))\right)/q, \qquad t_0 = \ln\left(Q_0/(1 + \exp(-t_0))\right)/q, \\ &t - t_0 = \ln\left((Q/Q_0)(1 + \exp(-t_0))/(1 + \exp(-t))\right)/q \geqq \ln(Q/Q_0)/q, \\ &t_{2,2} \geqq t_0 + \ln\left((Q/Q_0)/(1 - q + q(Q_0/Q)^{1/q})\right)/q =: C(\sigma). \qquad \text{Q.e.d.} \end{aligned}\right\} \tag{2.20}$$

3. Ein Einschluß mit exponentiell abnehmender Streifenbreite

Ein Ergebnis von G. Pólya [4], auf den Fall von Trinomen T angewandt, wobei das zugehörige Indikatordiagramm ein nichtausgearteter Dreiecksbereich ist, besagt, daß T für große $|z|$ drei Nullstellenfolgen besitzt, zu denen die Streifenmitten aus Satz 2.1 Asymptoten sind. Indem wir die bislang gewonnenen konstanten Streifenbreiten zu Streifen verbessern, die sich mit wachsendem $|z|$ exponentiell verjüngen, geben wir dem Pólyaschen Ergebnis für Trinome T eine quantitative Fassung. Für unsere jetzigen Ziele ist die Normierung (2.1) nicht mehr hilfreich, und wir setzen sie nicht länger voraus.

Satz 3.1: *Besitzt* conv (Q) *drei Eckpunkte, so liegen alle Nullstellen von* T *im krummlinig berandeten Streifen*

$$\left.\begin{aligned}
&S := S_a \cap S_b \cap S_c\,, \quad S_a := \big(W_a - \ln(A_b/A_a)\big)/(b-a)\,,\\
&W_a := \{w \in \mathbb{C} \mid w = u + iv,\ V_1(u) \leqq v \leqq V_2(u),\ u \in \mathbb{R}\}\,,\\
&V_1(u) := \big(V(u) - d_1(u)\big)/p_2\,, \quad V_2(u) := \big(V(u) + d_2(u)\big)/p_2\,,\\
&V(u) := (L + p_1 u - |u|)/p_2\,, \quad d_1(u) := -2\ln\big(1 - \exp(-|u|)\big)/p_2\,,\\
&d_2(u) := 2\ln\big(1 + \exp(-|u|)\big)/p_2\,,\\
&p := p_1 + ip_2 := (c - a + c - b)/(b - a)\,, \quad p_1, p_2 \in \mathbb{R}\,,\\
&L := \ln(|P|^2)\,, \quad P := A_c \exp\big(-p\ln(A_b/A_a)\big)/\sqrt{A_aA_b}\,.
\end{aligned}\right\} \tag{3.1}$$

Die asymptotische Streifenbreite von S_a *für* $|w| \to \infty$ *ist in der* w*-Ebene*

$$d(w) \approx 4\exp(-|w|)/p_2\,. \tag{3.2}$$

Vor dem Beweis zu vorstehendem Satz sei die nunmehr erreichte Nullstelleneinschließung anhand einiger Bilder illustriert. Wir gehen dabei bereits von der normierten Form

$$T(z) := 2\,\mathrm{ch}\,(z/2) - P\exp(pz/2) \tag{3.3}$$

aus. Wie schon bei Bild 1 sind jeweils das Indikatordiagramm und seine Ecken mitgezeichnet. Daneben sind noch die drei Asymptoten der Streifen angegeben. Die positive imaginäre Achse $i\mathbb{R}_+$ ist dabei jeweils die dritte Asymptote. Im Bild 2 ist $|P| = 1$ und $p = 2i$.

In den Bildern 3 bis 5 ist $p = 1 + i$ und $|P|$ variiert von $|P| = 1$ in Bild 3 über $|P| = 2$ in Bild 4 zu $|P| = 1/2$ in Bild 5. Man sieht, daß sich die Einschlußmenge mit wachsendem $|P|$ in Richtung von $i\mathrm{R}_+$ parallel verschiebt.

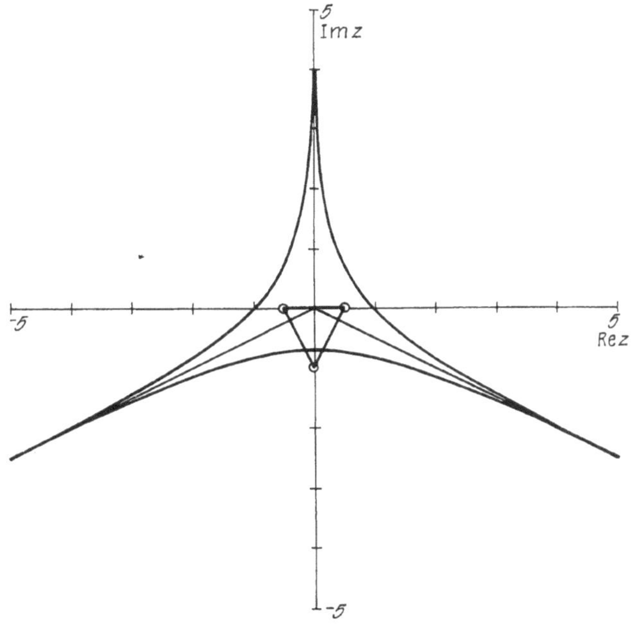

Bild 2. Der Einschluß von Satz 3.1 für $T = 2\,\mathrm{ch}\,(z/2) - \exp(iz/2)$

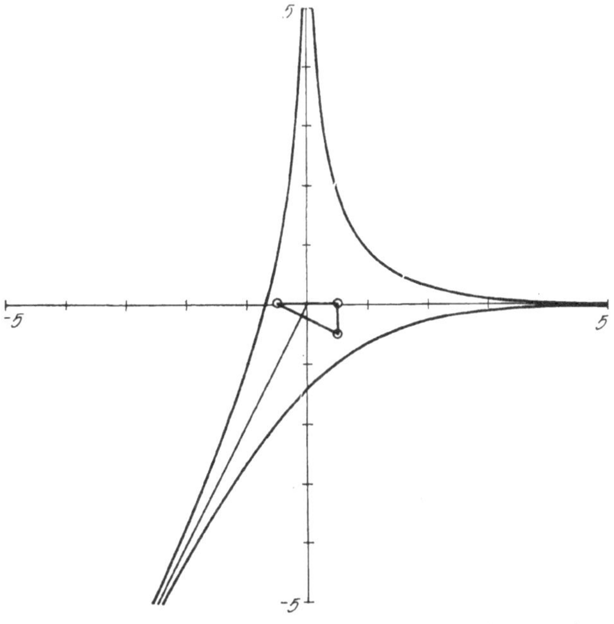

Bild 3. Der Einschluß von Satz 3.1 für $T = 2\,\mathrm{ch}\,(z/2) - \exp\big((1 + i)\,z/2\big)$

Bild 6 zeigt den Fall eines der reellen Achse nahen Punktes p, nämlich $p \doteq i/2$, $|P| = 1$. Rückt man mit p der reellen Achse noch näher, wie in Bild 7 mit $p = i/8$ und passend gewähltem $|P| = 4$, so wird klar, wie aus diesem Einschluß im Grenzübergang $\mathrm{Im}\,(p) \to 0$ der Fall des in Satz 2.2 beschriebenen Einschlusses erreicht wird. Da der Fall reeller Exponentenpunkte eine gewisse Eigenständigkeit in den Anwendungen besitzt, wurde von einer einheitlichen Formulierung der Sätze 2.2 und 3.1 abgesehen.

Wie im Beweis in (3.9) noch zu sehen sein wird, können die in den vorangehenden Bildern gezeigten Streifen als Bilder eines Streifens mit den Randkurven $g_1(u)$ und $g_2(u)$ in einer w-Ebene aufgefaßt werden. In Bild 8 ist eben-

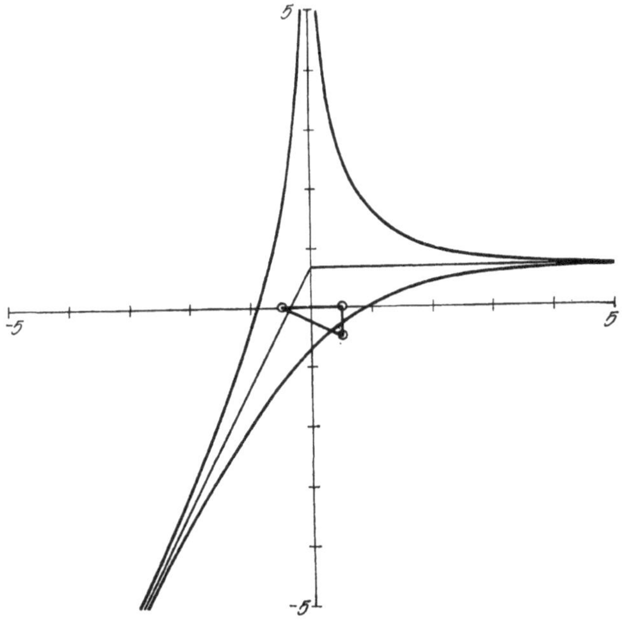

Bild 4. Der Einschluß von Satz 3.1 für $T = 2\,\mathrm{ch}\,(z/2) - \sqrt{2}\exp((1+i)\,z/2)$

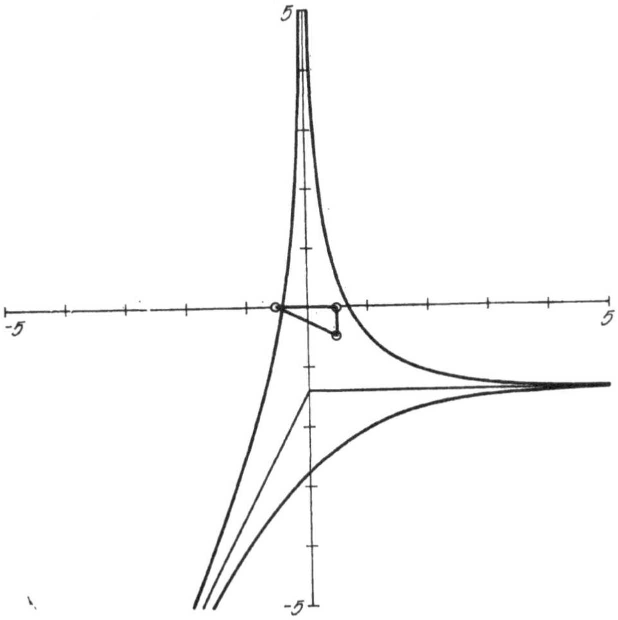

Bild 5. Der Einschluß von Satz 3.1 für $T = 2\,\mathrm{ch}\,(z/2) - \exp((1+i)\,z/2)/2$

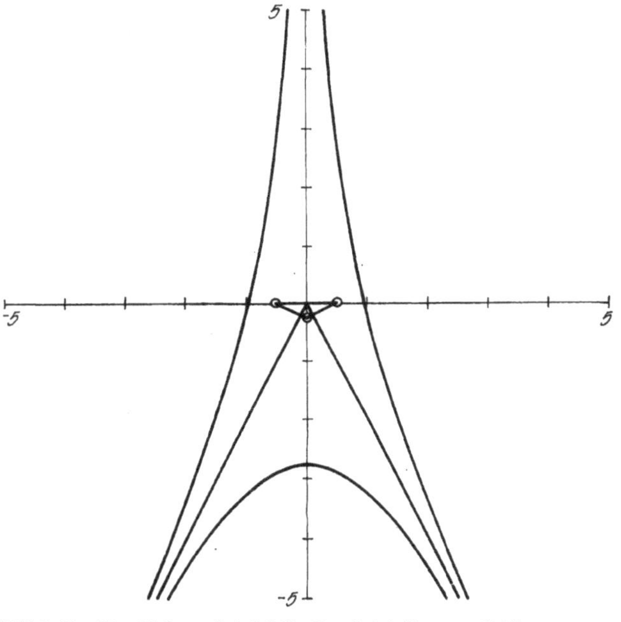

Bild 6. Der Einschluß von Satz 3.1 für $T = 2\,\mathrm{ch}\,(z/2) - \exp(iz/4)$

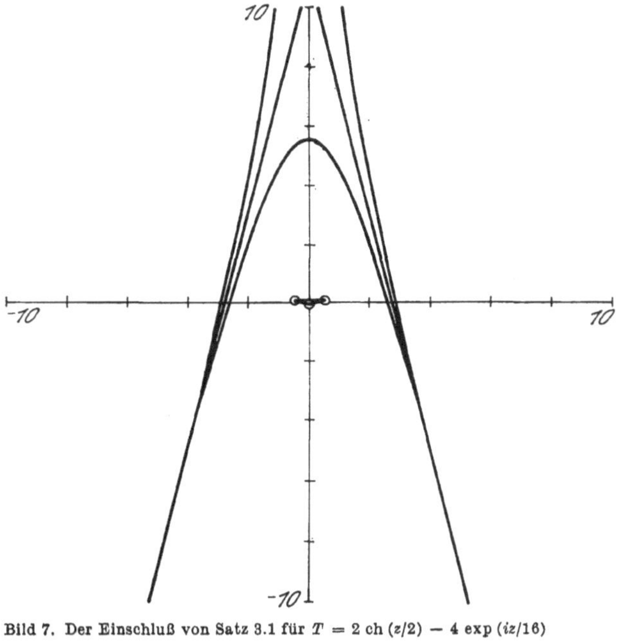

Bild 7. Der Einschluß von Satz 3.1 für $T = 2\,\mathrm{ch}\,(z/2) - 4\exp(iz/16)$

falls gezeigt, wie man aus den genaueren krummlinig berandeten Streifen wieder geradlinig berandete, wie den von Satz 2.1, zurückgewinnen kann.

Beweis: Wir wählen eine Ecke a und ihre Nachfolger in $V(Q)$ aus. Hinsichtlich der getroffenen Auswahl schreiben wir die Nullstellengleichung in der Form

$$A_a \exp(az) + A_b \exp(bz) = -A_c \exp(cz)\,. \tag{3.4}$$

Nach Multiplikation von (3.4) mit $\exp\big(-(a+b)\,z/2\big)/\sqrt{A_aA_b}$ erhalten wir

$$\sqrt{A_a/A_b}\exp\big((a-b)\,z/2\big) + \sqrt{A_b/A_a}\exp\big((b-a)\,z/2\big) = -A_c\exp\big((c-a+c-b)\,z/2\big)/\sqrt{A_aA_b}\,. \tag{3.5}$$

Mit

$$w = (b-a)\,z + \ln(A_b/A_a) \tag{3.6}$$

wird (3.6) in die w-Ebene übertragen, und man erhält dort

$$2\,\mathrm{ch}\,(w/2) = P \exp\,(pw/2)\,,$$
$$p := (c - a + c - b)/(b - a)\,, \qquad P := -A_c \exp\big(-\ln\,(A_b/A_a)\,p\big)/\sqrt{A_a A_b}\,. \tag{3.7}$$

Die Ähnlichkeitstransformation (3.6) führt die Eckpunkte a, b der z-Ebene in die Eckpunkte $-1/2$, $1/2$ der w-Ebene über. Die w-Ebene wird auf kartesische Koordinaten $w := u + iv$, $u, v \in \mathbb{R}$, bezogen. Nach Betragsbildung und Quadrieren in (3.7) bekommt man mit $c := \cos\,(v)$ und $\sigma := \mathrm{sgn}\,(u)$ sowie $p := p_1 + ip_2$ und $L := \ln\,(|P|^2)$

$$\exp\,(u) + 2c + \exp\,(-u) = |P|^2 \exp\big(\mathrm{Re}(pw)\big)\,. \tag{3.8}$$

Wir fassen ch(u) als Funktion von $|u|$ auf und erhalten nach Extremwertbildung bezüglich c

$$\begin{aligned} g_1(u) &:= 1 - \exp\,(-|u|) \leqq |P| \exp\big(\mathrm{Re}\,(pw)/2 - |u|/2\big) \leqq 1 + \exp\,(-|u|) =: g_2(u)\,, \\ 2 \ln\big(1 - \exp\,(-|u|)\big) &\leqq L + p_1 u - |u| + p_2 v \leqq 2 \ln\big(1 + \exp\,(-|u|)\big)\,. \end{aligned} \tag{3.9}$$

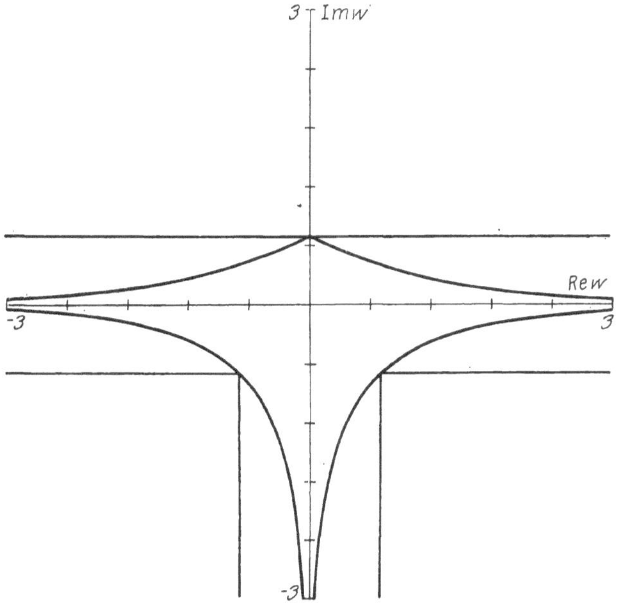

Bild 8. Das Urbild des Einschlusses von Satz 3.1 in der Hilfsebene w

Da b der Nachfolger von a ist, liegt c zur Linken des Vektors $b - a$, und es ist $p_2 > 0$. Auflösen der letzten Zeile in (3.9) nach v erbringt den Streifeneinschluß

$$\left.\begin{aligned} &V_1(u) := V(u) - d_1(u) \leqq v \leqq V(u) + d_2(u) =: V_2(u)\,, \\ &V(u) := (L + p_1 u - |u|)/p_2\,, \qquad d_1(u) := 2 \ln\big(1/(1 - \exp\,(-|u|))\big)/p_2\,, \\ &d_2(u) := 2 \ln\big(1 + \exp\,(-|u|)\big)/p_2\,. \end{aligned}\right\} \tag{3.10}$$

Durch (3.10) ist eine Umgebung von $\mathbb{R}_- \cup \mathbb{R}_+ \cup i\mathbb{R}_-$ in der w-Ebene bestimmt, deren Streifenbreite exponentiell mit $|w| \to \infty$ fällt. Die asymptotische Streifenbreite ist $\approx 4 \exp\,(-|w|)$. Rücktransformation des Einschlusses (3.10) für die Nullstellen des Trinoms in (3.7) mittels der Rücktransformation

$$z = \big(w - \ln\,(A_b/A_a)\big)/(b - a)\,, \tag{3.11}$$

zu (3.6), gefolgt von der Anwendung des Einschlusses auf alle drei a, gibt die im Satz festgehaltene Aussage. Q.e.d.

4. Ein zweiter krummlinig berandeter Einschluß

Wie aus dem Beweis in Satz 3.1 zu entnehmen war, liegen die Nullstellen von T auf einer Kurve, die zwischen den angegebenen Streifenrändern oszilliert. Es läßt sich zu jedem vorgegebenen Punkt auf dem Rand der Einschlußmenge ein Trinom angeben, das ihn als Nullstelle besitzt. Um eine weitere Einschließung zu erreichen, suchen wir nach Kurven, die die Randkurven des vorstehenden Einschlusses orthogonal oder wenigstens näherungsweise orthogonal durchsetzen.

Satz 4.1: *Die Nullstellen von T liegen im Streifensystem*

$$S := S_a \cap S_b \cap S_c\,, \qquad S_a := \big(W_a - \ln(A_b/A_a)\big)/(b-a)\,,$$
$$W_a := \{w \in \mathbb{C} \mid w := u + iv,\, u \in U,\, v \in \mathbb{R}\}\,,$$
$$U := \begin{cases} [U_1, U_2] \cup [U_3, U_4]\,, & p_1 < 1\,,\\ [U_5, U_6]\,, & p_1 > 1\,,\end{cases}$$
$$U_k := (p_2 v - L + d_k)/(1 + p_1) \leqq 0\,, \qquad k = 1, 2\,,$$
$$U_k := -(p_2 v - L + d_{5-k})/(1 - p_1) \geqq 0\,, \qquad k = 3, 4\,,$$
$$U_5 := (p_2 v - L + d_1)/\big(p_1 - \operatorname{sgn}(p_2 v - L + d_1)\big)\,, \quad U_6 := (p_2 v - L + d_2)/\big(p_1 - \operatorname{sgn}(p_2 v - L + d_2)\big)\,,$$
$$d_1 := H(-c)\ln\big(\sin^2(v)\big)\,, \qquad d_2 := \max\big(0, \ln(2 + 2\cos(v))\big)\,.$$

Dabei sind L und p wie im Satz 3.1 definiert.

Falls für $p_1 < 1$ die Ungleichung $U_2 \leqq 0 \leqq U_3$ verletzt ist, aber $U_1 \leqq 0 \leqq U_4$ erfüllt ist, ist $[U_1, U_4]$ das Lösungsintervall. Ist die zweite Ungleichung auch noch verletzt, so ist die Gerade Im $(w) = v$ *nullstellenfrei.*

Die Bilder 9 bis 15 zeigen die Einschlüsse zum vorangehenden Satz für das Trinom (3.7) und den Rand des zugehörigen Indikatordiagramms.

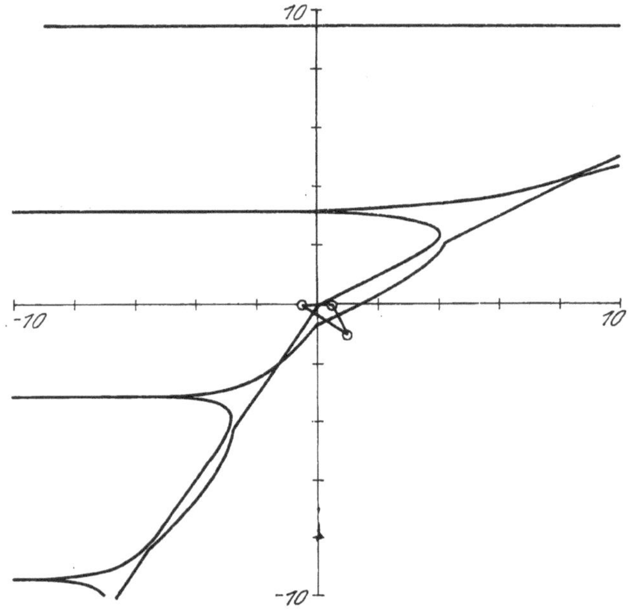

Bild 9. Der Einschluß von Satz 4.1 für $T = 2\,\mathrm{ch}\,(z/2) - \exp((1 + i)z)$

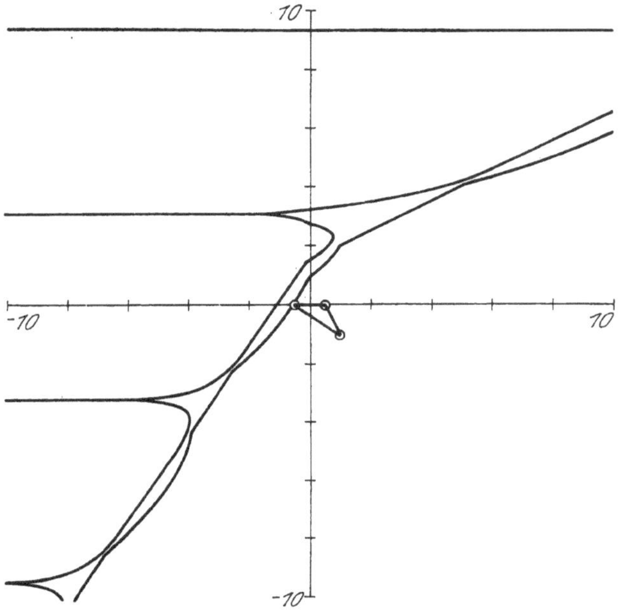

Bild 10. Der Einschluß von Satz 4.1 für $T = 2\,\mathrm{ch}\,(z/2) - 5\exp((1 + i)z)$

In Bild 9 ist $|P| = 1$ und $p = 2 + 2i$. Änderung von $|P|$ äußert sich im wesentlichen durch eine Parallelverschiebung des Einschlusses, wie Bild 10 mit $|P| = 5$ und ungeändertem p zeigt. Auf die in der Nähe der positiven imaginären Achse gelegenen Nullstellen hat eine Änderung von p wenig Einfluß, vgl. Bild 11, in dem $|P| = 1$ und $p = (3 + i)/2$ ist. Für Re $(p) < 1$ ändert sich die Geometrie der Einschlüsse. In den Bildern 12 bis 14 ist jeweils $p = i$ und $|P| = 0.1, 1, 10$ in dieser Reihenfolge. Ein zur imaginären Achse unsymmetrisches Randdreieck des Indikatordiagramms zeigt Bild 15 mit $|P| = 0.2$ und $p = (1 + 4i)/2$. Die Einschlüsse der Bilder 9 bis 15 können als Bilder der in Bild 16 gezeigten Menge angesehen werden. Teile der reellen Achse gehören zur unteren Berandung.

Beweis: Wir benutzen den Anfang des Beweises zu Satz 3.1 bis zur Formel (3.8),

$$\exp(u) + 2c + \exp(-u) = |P|^2 \exp\big(\mathrm{Re}\,(pw)\big)\,. \tag{4.5}$$

Mit $L := 2\ln(|P|)$ und $p =: p_1 + ip_2$ sowie $s := \sin(v)$ erhalten wir nach Logarithmen von (4.5)

$$\ln\big(s^2 + (c + \exp(-|u|))^2\big) = L + p_1 u - |u| - p_2 v\,. \tag{4.6}$$

Die linke Seite von (4.6) schätzen wir durch Funktionen ab, die nur von v abhängen. Geometrisch bedeutet dies, daß wir obere und untere Enveloppe der linken Seite von (4.6), verstanden als Funktionenschar mit dem Scharparameter u, ermitteln.

$$\left.\begin{aligned} F(u) := \ln\big(s^2 + (c + \exp(-|u|))^2\big) &\leqq \max\{F(u) : u \in \mathbb{R}\} = \ln_+(2 + 2c)\,,\\ &\geqq \min\{F(u) : u \in \mathbb{R}\} = H(-c)\ln(s^2)\,,\\ \ln_+(t) := \max\big(0, \ln(t)\big)\,, &\quad t \in \mathbb{R}_+\,. \end{aligned}\right\} \tag{4.7}$$

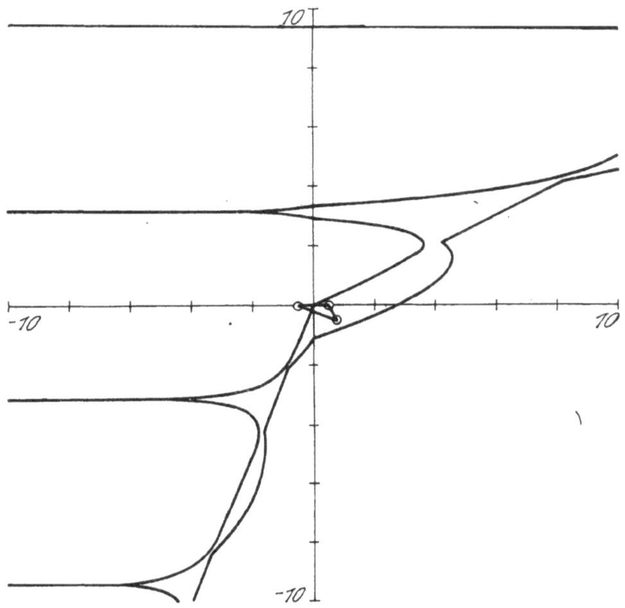

Bild 11. Der Einschluß von Satz 4.1 für $T = 2\,\mathrm{ch}\,(z/2) - \exp\,((3 + i)\,z/4)$

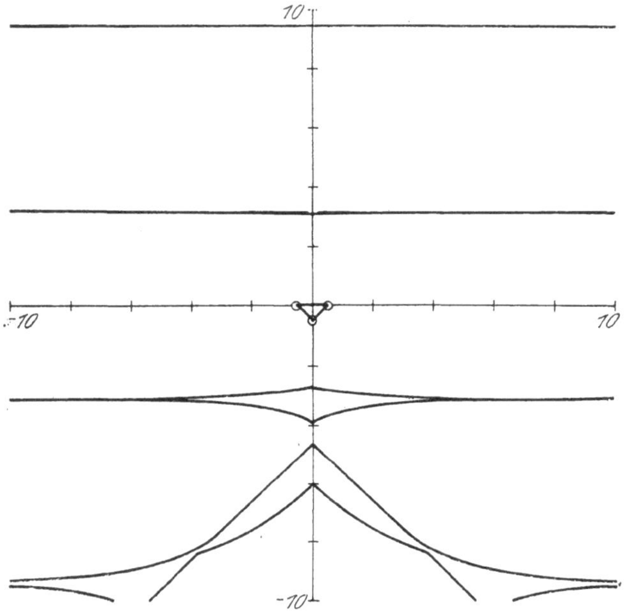

Bild 12. Der Einschluß von Satz 4.1 für $T = 2\,\mathrm{ch}\,(z/2) - \exp\,(iz/2)/10$

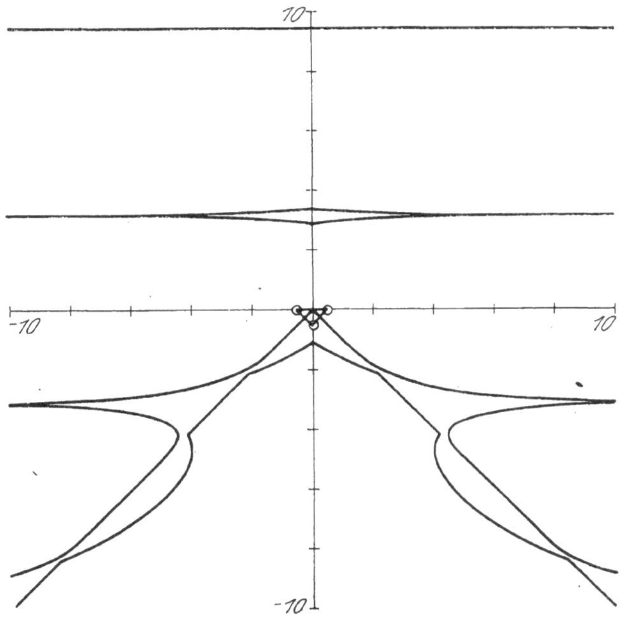

Bild 13. Der Einschluß von Satz 4.1 für $T = 2\,\mathrm{ch}\,(z/2) - \exp\,(iz/2)$

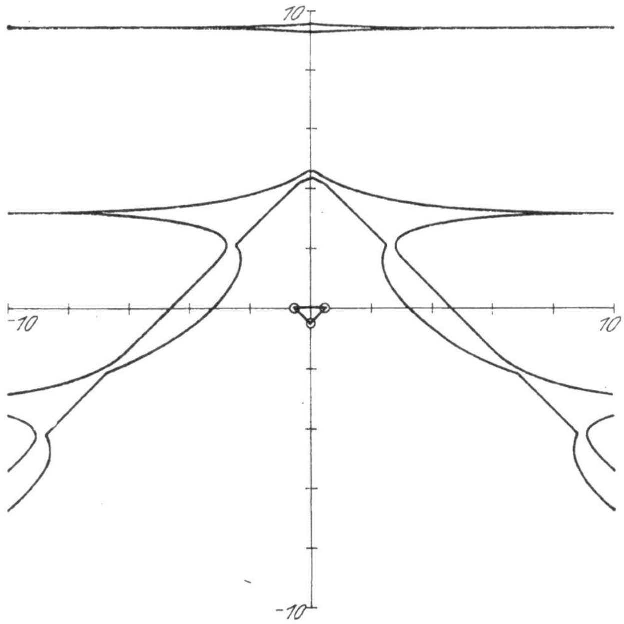

Bild 14. Der Einschluß von Satz 4.1 für $T = 2\,\mathrm{ch}\,(z/2) - 10\,\exp\,(iz/2)$

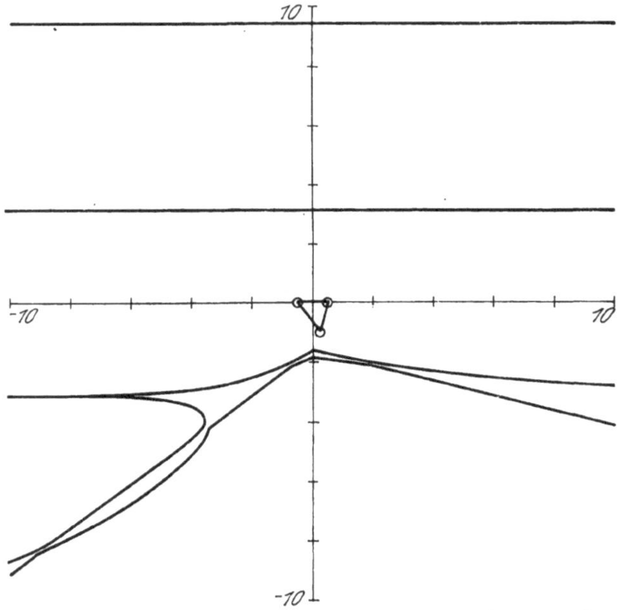

Bild 15. Der Einschluß von Satz 4.1 für $T = 2\,\mathrm{ch}\,(z/2) - \exp\left((1 + 4i)\,z/4\right)/5$

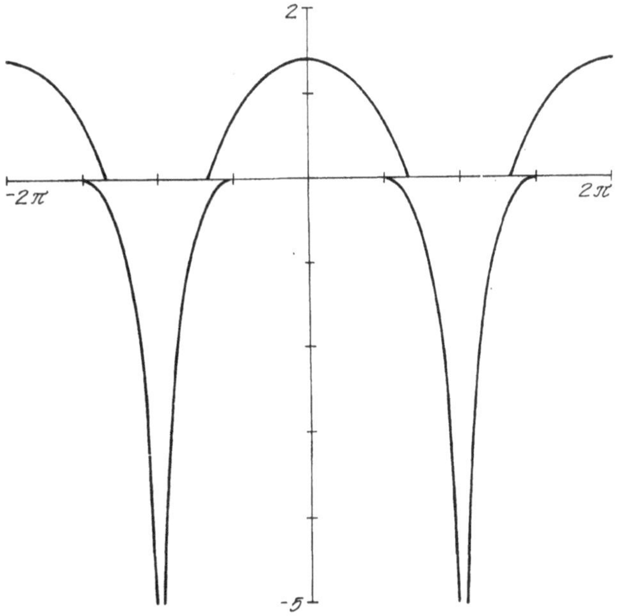

Bild 16. Das Urbild des Einschlusses von Satz 5.1 in der Hilfsebene w

Einsetzen von (4.7) in (4.6) führt auf

$$\left.\begin{aligned} &U(v) - d_1(v) \leqq p_1 u - |u| \leqq U(v) + d_2(v)\,, \\ &U(v) := p_2 v - L\,, \qquad d_1(v) := -H(-c)\ln(s^2)\,, \qquad d_2(v) := \ln_+(2 + 2c)\,. \end{aligned}\right\} \tag{4.8}$$

Durch (4.8) veranlaßt, betrachten wir mit $p, t \in \mathbb{R}$ die Gleichung für u

$$t = pu - |u|\,, \tag{4.9}$$

deren Lösungszweige u durch

$$u = u(t) := \begin{cases} -t\sigma/(1 - p\sigma)\,, & p < 1\,, \quad t \leqq 0\,, \quad \sigma \in \{-1, 1\}\,, \\ -\infty\,, & p \leqq 1\,, \quad t > 0\,, \\ t/\left(p - \operatorname{sgn}(t)\right)\,, & p > 1\,, \\ \mathbb{R}_+\,, & p = 1\,, \quad t = 0\,, \\ t/2\,, & p = 1\,, \quad t < 0\,, \end{cases} \tag{4.10}$$

gegeben sind. Für $p < 1$ und $t \leqq 0$ gibt es genau zwei Lösungen. Die Lösungsdarstellung in (4.10) ist so eingerichtet daß u mit wachsendem σ wächst. Für $p > 1$ gibt es nur eine Lösung. Für $p \leqq 1$ und $t > 0$ hat (4.9) keine Lösung Das heißt, daß für v, für die die linke Schranke in (4.8) positiv ausfällt, auf der Geraden $\operatorname{Im}(w) = v$ keine Nullstellen liegen. Wir fassen dies in die Form $u = -\infty$.

Nach Verwenden der äußeren Glieder der Ungleichung in (4.8) als Argumente von $u(t)$ bekommen wir die im Satz angegebenen Intervallgrenzen. Q.e.d.

5. Ein Kreisscheibeneinschluß für fast alle Nullstellen

Die Einschlüsse der Sätze 3.1 und 4.1 sind gleichzeitig gültig. Also liegen die Nullstellen von T im Durchschnitt beider Einschlüsse. Die Einschlußmengen sind fast alle krummlinig berandete Rechtecksbereiche. Diese kann man noch zu bequemeren Rechtecken vergrößern. Wir lassen noch einen größeren Genauigkeitsverlust zu und suchen nach Kreisscheibeneinschlüssen. Eckige Klammern bezeichnen die Funktion des größten ganzzahligen Teiles.

Im weiteren sei

$$K(m, r) := \{z \in \mathbb{C} \mid |z - m| \leqq r\} \tag{5.1}$$

die Kreisscheibe mit dem Mittelpunkt $m \in \mathbb{C}$ und dem Radius $r \in \mathbb{R}_+$.

Satz 5.1: *Die Kreisscheiben der Kreisscheibenfolge*

$$\left.\begin{aligned} &C := C_a \cup C_b \cup C_c\,, \qquad C_a := \left(K_a - \ln(A_b/A_a)\right)/(b - a)\,, \qquad a \in Q\,, \\ &K_a := \bigcup_{k \in \mathbb{N}_a} K_{a,k}\,, \qquad K_{a,k} := K(m_{a,k}, r_{a,k})\,, \\ &m_{a,k} := i\pi(1 + 2k)\,, \quad r_{a,k} := \pi \exp\left((\ln(|P_a|^2/4) + |p_a|\,\pi - p_{a,2}(1 + 2k)\,\pi)/2\right), \quad \mathbb{N}_a := k_{a,0} + \mathbb{N}_0\,, \\ &k_{a,0} := -\left[1/2 - (|p_a| + \ln(|P_a|^2/4)/\pi)\right], \qquad p_a := p_{a,1} + ip_{a,2} := (2c - a - b)/(b - a)\,, \\ &P_a := -A_c \exp\left(-p_a \ln(A_b/A_a)\right)/\sqrt{A_a A_b}\,, \end{aligned}\right\} \tag{5.2}$$

enthalten eine Nullstelle von T. Wird $\mathbb{N}_a$ *mit*

$$k_{a,1} := -[-\{(\ln(|P_a|(1+\exp(-p_2\pi))/4)/\pi + |p|/2)/p_2 - 1/2\}],$$

anstelle von $k_{a,0}$ *definiert, so trennen die Kreisscheiben mit festem* $a \in Q$ *einander und enthalten genau eine Nullstelle von T. Die Punkte a, b, c treten in dieser Reihenfolge in V auf.*

Beweis: Wir folgen dem Beweis von Satz 3.1 bis zur Formel (3.7),

$$2\,\mathrm{ch}\,(w/2) = P \exp(pw/2)\,. \tag{5.3}$$

Nach Voraussetzung hat conv (Q) drei Ecken, somit gilt $p_2 := \mathrm{Im}\,(p) > 0$, wobei $p := p_a$ ist. Der Betrag der längs der positiven imaginären w-Achse genommenen rechten Seite $|P| \exp(-p_2 v/2)$ von (5.3) ist für große positive $p_2 v$ genügend klein. Für betragsgroße w nahe $i\mathbb{R}_+$ kann die rechte Seite von (5.3) als kleine Strömung der homogenen Gleichung $P := P_a = 0$, zu (5.3) aufgefaßt werden. Dementsprechend setzen wir für die nahe $i\mathbb{R}_+$ vermuteten (und durch frühere Sätze schon eingeschlossenen) Lösungen von (5.3)

$$w = i\pi(1+2k) + s\,, \qquad k \in \mathbb{Z}\,, \tag{5.4}$$

an. Mit dem Ansatz (5.4) schreibt sich (5.3) als

$$\mathrm{sh}\,(s/2) = P_k \exp(ps/2)\,, \qquad P_k := i(-1)^{k+1} P \exp(ip\pi(1/2+k))/2\,. \tag{5.5}$$

Ist C eine geschlossene, sich selbst nicht schneidende Kurve der s-Ebene, auf der

$$|\mathrm{sh}\,(s/2)| > |P_k \exp(ps/2)| \tag{5.6}$$

erfüllt ist, so liegen nach dem Satz von Rouché aus der Funktionstheorie im Inneren von C ebensoviele Lösungen von (5.5) wie (5.5) bei verschwindender rechter Seite dort hat. Die Klasse der zugelassenen Kurven C wurde schon auf Kreise festgesetzt. Wir gehen noch weiter und betrachten anstatt aller Kreise, die nur die Nullstelle $s = 0$ enthalten, nur die Kreise mit $s = 0$ als Mittelpunkt und dem durch $r \leqq \pi$ beschränkten Radius. Unter dieser Radienbeschränkung gilt

$$|\mathrm{sh}\,(s/2)| \geqq \sin(r/2)\,, \qquad |\exp(ps/2)| \leqq \exp(|p|\,\pi/2)\,. \tag{5.7}$$

Die Ungleichung (5.6) besteht, wenn sogar

$$\sin(r/2) \geqq r/\pi > |P_k| \exp(|p|\,\pi/2)\,, \tag{5.8}$$

erfüllt ist. Dies ist unter der getroffenen Radienbeschränkung für die $k \in \mathbb{Z}$ mit

$$k \geqq k_0' := (|p| + \ln(|P|^2)/\pi)/(2p_2) - 1/2 \tag{5.9}$$

erfüllbar. Das kleinste k, das (5.9) löst, ist

$$k_0 := -[-k_0']\,. \tag{5.10}$$

Ist $k \in k_0 + \mathbb{N}_0$, so wird (5.8) durch

$$r > r_k\,, \qquad r_k := \pi \exp\{(\ln(|P|^2/4) + (|p| - p_2(1+2k)\,\pi))/2\} \tag{5.11}$$

gelöst. Da sh $(s/2)$ in jeder Kreisscheibe $|s| \leqq r$ eine Nullstelle hat, trifft dies auf die Lösungen der Gleichung (5.5) und für die abgeschlossenen Kreisscheiben mit gleichem Mittelpunkt und den in (5.11) gegebenen Radien r_k zu.

Zwei benachbarte Kreisscheiben mit den Radien a und b aus (5.11) haben den Mittelpunktsabstand 2π. Deswegen liegen sie für $a + b < \pi$ getrennt. Die Trennungsbedingung für zwei aufeinanderfolgende Kreisscheiben K_k und K_{k+1} dieser Art ist somit

$$r_k + r_{k+1} < 2\pi\,. \tag{5.12}$$

Sie ist für

$$k_1 := -[-\{\ln(|P|(1+\exp(-p_2\pi))/4)/\pi + |p|/2)\,p_2 - 1/2\}]\,, \tag{5.13}$$

erfüllt. Q. e.d.

In Tabelle 1 sind die Radien (als $3D$-Zahlen) der ersten einschließenden Kreisscheiben für das Trinom $T = 2\,\mathrm{ch}\,(z/2) - 10 \exp((1+i)\,z)$ angegeben. Zur Lage der ersten beiden dieser Nullstellen vergleiche Bild 10.

Tabelle 1. Die Radien der ersten einschließenden Kreisscheiben nach Satz 5.1 für $T = 2\,\mathrm{ch}\,(z/2) - 10 \exp((1+i)\,z)$

k	r	k	r	k	r
1	1.08D-1	4	7.02D-10	7	4.57D-18
2	2.01D-4	5	1.31D-12	8	8.54D-21
3	3.76D-6	6	2.45D-15	9	1.59D-23

$k_0 = 1$, $k_1 = 1$, $m_k = i\pi(1+2k)$

6. Diskussion

Wie einleitend erwähnt, können gewöhnliche Trinome in der Variablen $z \in \mathbb{C}$

$$T(z) := Az^a + Bz^b + Cz^c \,, \tag{6.1}$$

mit $a < b < c$, $a, b, c \in \mathbb{N}_0$ sowie komplexen nicht verschwindenden Koeffizienten $A, B, C \in \mathbb{C}$ als Exponentialpolynome in der Variablen w, die durch $z := \exp(w)$ definiert wird, angesehen werden. Die Exponentenpunkte liegen dabei in $\mathbb{R}_+$. Wir wollen nachfolgend stets von den eventuellen trivialen Nullstellen $z = 0$ absehen.

Die Trinome T, die wir auch als lakunäre Polynome auffassen können, sollen uns als Test- und Demonstrationsbeispiel für die gewonnenen Resultate dienen. Auf eine Gegenüberstellung mit bekannten Nullstelleneinschlüssen für Trinome T verzichten wir, weisen aber auf die Monographien von M. Marden [3], J. L. Walsh [6] sowie auf W. Specht [5], J. Dieudonné [2] und M. Biernacki [1] hin. Der Trinomfall eignet sich als Testfall, da er einerseits als geläufig angesehen werden darf und andererseits im Fall kollinearer Exponentenpunkte die Nullstellen am schwersten zu trennen sind.

Satz 2.1 liefert unter der Koeffizientennormierung (2.1) den Kreisringeinschluß

$$\left.\begin{aligned} &R_1 \leqq |z| \leqq R_4\,, \quad R_1 := \big(|A|/(1-|A|)\big)^{1/(n_a-a)}\,, \quad R_4 := \big(|C|/(1-|C|)\big)^{1/(n_c-c)}\,, \\ &n_a := \begin{cases} b, & |A| \leqq 1/2 \\ c, & |A| \geqq 1/2 \end{cases}, \quad n_c := \begin{cases} b, & |C| \leqq 1/2 \\ a, & |C| \geqq 1/2 \end{cases}, \end{aligned}\right\} \tag{6.2}$$

für die nichttrivialen Nullstellen von T. Für $|B| > 1/2$ verzweigt der Kreisring (6.2) in zwei disjunkte Kreisringe

$$|z| \in [R_1, R_2] \cup [R_3, R_4]\,, \qquad R_2 := \big((1-|B|)/|B|\big)^{1/(b-a)}\,, \qquad R_3 := \big(|B|/(1-|B|)\big)^{1/(c-b)}\,. \tag{6.3}$$

Dabei sind R_1 und R_2 aus (6.2) zu nehmen. Für $|B| \to 1$ müssen die nichttrivialen Nullstellen zum Ursprung $z = 0$ oder nach dem Unendlichen ausweichen. Die Radien R_1 bis R_4 qualifizieren das Wachstum des für $|B| > 1/2$ sich einstellenden offenen nullstellenfreien Kreisrings zwischen den beiden Ringen, in denen Nullstellen möglich sind, zur punktierten Ebene $\mathbb{C} \setminus \{0\}$. Für $|B| \to 0$ bewegen sich die nichttrivialen Nullstellen auf den Kreis $|z| = |A/C|^{1/(c-a)}$ zu, dessen Radius ersichtlich der gemeinsame Grenzwert von R_1 und R_4 für $|B| \to 0$ ist.

Im Sinne von Satz 2.2 wird die Nullstellengleichung $T = 0$ in der Form

$$\left.\begin{aligned} &\sqrt{w} + 1/\sqrt{w} = P w^{p/2}\,, \\ &w := z^{c-a} C/A\,, \qquad p := (b-a+b-c)/(c-a)\,, \qquad P := -B(A/C)^{p/2}/\sqrt{AC} \end{aligned}\right\} \tag{6.4}$$

geschrieben. Wir benutzen den Diskussionsteil, um nochmal auf die methodisch tragende Minimax-Zerlegung (für den Fall zweier reeller Summanden)

$$a + b = \max(a, b) + \min(a, b) \tag{6.5}$$

hinzuweisen. Gemäß (6.5) führen wir $\sigma := \operatorname{sgn}(r-1)$, $r := |w|$, ein und erhalten aus (6.4)

$$\left.\begin{aligned} &|1/\sqrt{r} - \sqrt{r}| \leqq |P|\, r^{p/2} \leqq 1/\sqrt{r} + \sqrt{r}\,, \\ &r^{\sigma/2} - r^{-\sigma/2} \leqq |P|\, r^{\sigma(\sigma p/2)} \leqq r^{\sigma/2} + r^{-\sigma/2}\,, \qquad 1 - r^{-\sigma} \leqq |P|\, r^{-\sigma(1-\sigma p)/2} \leqq 1 + r^{-\sigma}\,. \end{aligned}\right\} \tag{6.6}$$

Die Substitution $Q := |P|$, $q := (1-\sigma p)/2$, $r^{-\sigma} := \exp(-t)$, also $t := |\ln(r)|$, führt auf

$$1 - \exp(-t) \leqq Q \exp(-qt) \leqq 1 + \exp(-t) \tag{6.7}$$

und also bis in den Beweis zu Satz 2.2 hinein. Dessen Anwendung liefert die radiale Einschließung

$$|w| \in \begin{cases} [R_1, R_4] & , \quad |P| \leqq P_0\,, \\ [R_1, R_2] \cup [R_3, R_4]\,, & \quad |P| \geqq P_0\,; \end{cases} \tag{6.8}$$

$$R_1 := \big(|P| + \min(1, |P|^{-1/r_0})\big)^{-2/(1+p)}\,,$$

$$R_2 := \begin{cases} \big(1 + ((1+r_0)/|P|)^{2/(1+p)}\big)/|P|\big)^{2/(1+p)}\,, & p \leqq 0\,, \quad P_0 \leqq |P|\,, \\ 1 + (r_0 - 1)(2 - |P|)/(2 - P_0) & , \quad p \geqq 0\,, \quad P_0 \leqq |P| \leqq 2\,, \\ \big((1 + (2/|P|)^{2/(1+p)})/|P|\big)^{2/(1+p)} & , \quad p \geqq 0\,, \quad 2 \leqq |P|\,; \end{cases}$$

$$R_3 := \begin{cases} r_0 + (1 - r_0)(|P| - P_0)/(2 - P_0) & , \quad p \leqq 0\,, \quad P_0 \leqq |P| \leqq 2\,, \\ \big(|P|/(1 + (2/|P|)^{2/(1-p)})\big)^{2/(1-p)} & , \quad p \leqq 0\,, \quad 2 \leqq |P|\,, \\ \big(|P|/(1 + ((1 + 1/r_0)/|P|)^{2/(1-p)})\big)^{2/(1-p)}, & \quad p \geqq 0\,, \quad P_0 \leqq |P|\,; \end{cases}$$

$$R_4 := \big(|P| + \min(1, |P|^{-r_0})\big)^{2/(1-p)}\,;$$

$$P_0 := (\sqrt{r_0} + 1/\sqrt{r_0})\, r_0^{(1-r_0)/(2+2r_0)}\,, \qquad r_0 := (1+p)/(1-p)\,.$$

Wir verfolgen die Entwicklung der Einschlußmenge für wachsende $|P|$. Für $|P| = 0$ ist der Kreisring $[R_1, R_4]$ zum Einheitskreis geschrumpft. Der Ring wächst beiderseits des Einheitskreises bis zum Verzweigungswert $|P| = P_0$. Für $|P| > P_0$ hat sich der Kreisring in zwei Kreisringe verzweigt, die zu beiden Seiten des Verzweigungsradius r_0 nach dem Ursprung bzw. nach dem Unendlichen für weiter anwachsende $|P|$ streben. Dabei nimmt die Breite beider Kreisringe wieder bis auf Null ab. Die Breite der Einschlußmenge ist für $|P| = P_0$ am größten.

Die schon erhaltene radiale Einschließung der Nullstellen ignorieren wir und ziehen Satz 4.1 zur Nullstelleneinschließung heran. Ohne (6.8) als a-priori-Einschließung zu benutzen, liefert er die Einschließung

$$\left.\begin{aligned}
&|w| \in [\min(1, R_1), \min(1, R_2)] \cup [\max(1, R_3), \max(1, R_4)],\\
&R_1 := \left((s^2 + H(c)\,c^2)/|P|^2\right)^{1/(1+p)}, \qquad R_2 := \left((1 + (1 + 2c)_+)/|P|^2\right)^{1/(1+p)},\\
&R_3 := \left((1 + (1 + 2c)_+)/|P|^2\right)^{-1/(1-p)}, \qquad R_4 := \left((s^2 + H(c)\,c^2)/|P|^2\right)^{-1/(1-p)},\\
&s := \sin(\varphi), \quad c := \cos(\varphi), \qquad (t)_+ := \max(0, t), \quad t \in \mathbb{R},\\
&\varphi := (c - a)\arg(z) + \arg(C/A).
\end{aligned}\right\} \tag{6.9}$$

In Tabelle 2 sind für $c - a = 6$, $C/A = 1$, $p = 1/3$ und $|P| = 0.5(0.5)\,2.5$ die durch (6.8) gelieferten Radien für den Einschluß der Nullstellen in der z-Ebene zusammengestellt und mit den exakten Radien verglichen. Außerdem sind die Breiten der Kreisringe aufgeführt. Unter den exakten Radien, die in Tabelle 2 mit dem Index ex versehen sind, werden die Lösungen von (6.6) mit Gleichheitszeichen verstanden.

Tabelle 2. Einschlußradien für $T(z) = 1 - Pz^4 + z^6$, $|P| = .5(.5)\,2.5$ nach (6.8)
(In die z-Ebene umgerechnete Kreisringradien nach (6.8))

$\lvert P\rvert$	R_1 / $R_{1,\mathrm{ex}}$	R_2 / $R_{2,\mathrm{ex}}$	R_3 / $R_{3,\mathrm{ex}}$	D_{32} / $D_{32,\mathrm{ex}}$	R_4 / $R_{4,\mathrm{ex}}$	D_{41} / $D_{41,\mathrm{ex}}$	B / B_{ex}
0.5	0.904	—	—	—	1.225	0.321	0.321
	0.926	—	—	—	1.094	0.168	0.168
1.0	0.841	—	—	—	1.414	0.573	0.573
	0.869	—	—	—	1.211	0.342	0.342
1.5	0.811	—	—	—	1.394	0.583	0.583
	0.823	—	—	—	1.344	0.521	0.521
2.0	0.780	1.000	1.186	0.186	1.500	0.720	0.534
	0.786	1.000	1.272	0.272	1.489	0.703	0.431
2.5	0.752	0.910	1.434	0.524	1.632	0.879	0.355
	0.756	0.870	1.521	0.651	1.626	0.870	0.219

$p = 1/3$, $r_0 = 1.122$, $P_0 = 1.890$, $c - a = 6$, $C/A = 1$, $D_{32} := R_3 - R_2$, $D_{41} := R_4 - R_1$, $B := D_{41} - D_{32}$

Die Einschließung durch (6.9) ist geometrisch recht gestaltreich. Es ist daher angebracht, sie eingehend zu beschreiben und am Beispiel darzustellen. Die $R_k = R_k(\alpha)$, $k = 1(1)\,4$, sind 2π-periodisch stetige ($|P| > 0$) positive Funktionen des Winkels $\alpha := \arg(z)$. Nur R_4 besitzt $c - a$ Pole. Die R_k beschreiben die Randkurven der Einschlußmenge in Polarkoordinaten $r := |z|$ und α. Bei festen anderen Parametern können, abhängig von $|P|$, 2 bis 4 geschlossene Randkurven auftreten. R_4 schließt sich über das Unendliche. Um unter diesen Umständen zu einer passenden Darstellung des Randes zu kommen, wurde die Einschlußmenge in der w-Ebene in zwei Komponenten zerlegt. Sie bestehen in den beiden Teilmengen der Einschlußmenge, die in $|w| \leqq 1$ und $|w| \geqq 1$ liegen. Die Vereinigung der Einschränkung der Kurven R_k auf beide Kreisbereiche ergibt die Ränder der Einschlußmenge.

Die Kurve R_4 verläuft in der w-Ebene längs des Kreises $|w| = |P|^{2/(1-p)} := R_5$ oder im Äußeren dieses Kreises. Für $|P| \geqq 1$ wird sie daher nicht eingeschränkt. Für die Winkel α mit $s = 0$ hat R_4 Pole. Die Nullstellen liegen im Innengebiet zu R_4, das den Ursprung enthält. Weiter liegen die Nullstellen im Außengebiet zu R_1. Diese Kurve verläuft längs des Kreises $|w| = |P|^{-2(1+p)} = R_6$ oder innerhalb des Kreises.

Der Ursprung wird $c - a$ mal in den Richtungen α, für die R_4 Pole hat, erreicht. Für $|P| < 1$ treten R_2 und R_3 nicht auf. Für $1 < |P| < 2$ treten im Bereich zwischen R_1 und R_4 nullstellenfreie „Inseln" auf. Es sind $c - a$ Stück, und sie werden durch die Einschränkungen der Kurven R_2 und R_3 berandet. Für weiter wachsende $|P|$ wachsen die Inseln und vereinigen sich für $|P| > 2$ zu einem zusammenhängenden nullstellenfreien Streifen, der im Ring $R_5 < |w| < R_6$ liegt. Für $c - a = 6$, $C/A = 1$ sowie $p = 1/3$ und $|P| = 0.5(0.5)\,1.5$ verfolgen wir die Entwicklung der Einschlußmenge in der z-Ebene in den Bildern 17 bis 19 in dieser Reihenfolge. Für $|P| = 0$ liefert (6.9) die Strahlen, auf denen die Nullstellen liegen. Zusammen mit der orthogonalen radialen Einschließung (6.8) erhalten wir die exakte Lage der Nullstellen. Für wachsende $|P|$ weiten sich die Strahlen auf, wie Bild 17 zeigt. Im Bild 18 für $|P| = 1.5$ haben sich die Strahlen soweit erweitert, daß sie miteinander Kontakt bekommen haben und gleichzeitig nullstellenfreie Inseln entstanden sind. Im Bild 18 für $|P| = 2.5$ haben sich die Inseln zu einem welligen Ring vereinigt. Für weiter wachsende $|P|$ zieht sich der innere Ring auf den Ursprung zusammen, und der äußere Ring weicht nach dem Unendlichen aus.

Für $|P| < 1$ haben sich die einschließenden Strahlenumgebungen noch nicht vereinigt. Zusammen mit der radialen Einschließung (6.8) entstehen $c - a$ getrennt liegende krummlinige Vierecke, die alle $c - a$ nichttrivialen Nullstellen einschließen. Für $|P| > 2$ hat die Einschließung (6.9) ebenfalls einen radialen Charakter, wenn man von den schmalen nach dem Ursprung und nach Unendlich laufenden Streifen absieht. Diese können auf einfache Weise als unrealistisch erkannt werden. Ob die gleichzeitige Verwendung von (6.8) und (6.9) noch eine Trennung ermöglicht, hängt von der Größe der Radien R_1 bis R_4 aus (6.8) und der zuvor definierten Radien R_5 und R_6 aus (6.9) ab. Für $R_1 < R_5 < R_2$ und $R_3 < R_6 < R_4$ werden keine Nullstellen in den beiden Kreisringen getrennt. Dies ist nach (6.8) und (6.9) erfüllt.

Immerhin wird die Einschließung von (6.8) durch (6.9) noch verbessert, wenn auch für $|P| > 2$ nicht mehr so drastisch, daß die Nullstellen getrennt werden. Die Vorgehensweise gestattet es, ähnliche Einschlüsse für allgemeine Exponentialpolynome zu gewinnen. Über die erhaltenen Ergebnisse soll zu anderer Gelegenheit berichtet werden.

Mittels der gegebenen Sätze lassen sich beispielsweise Mengen im Parameterraum, der von den Koeffizienten und Argumentverschiebungen aufgespannt wird, angeben, indem die zugrunde liegende Differenzengleichung asymptotisch stabil ist, vgl. etwa MELVIN [7], AVELLAR und HALE [8].

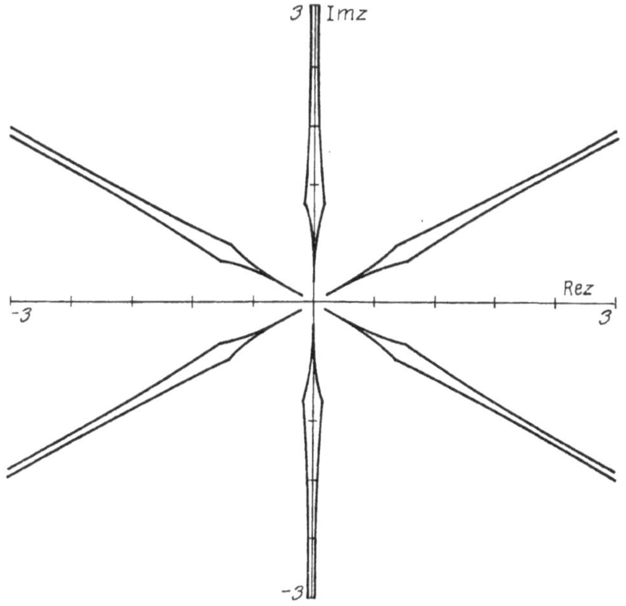

Bild 17. Der Einschluß für $T(z) = 1 - \varepsilon z^4/2 + z^6$, $|\varepsilon| = 1$ nach (6.9)

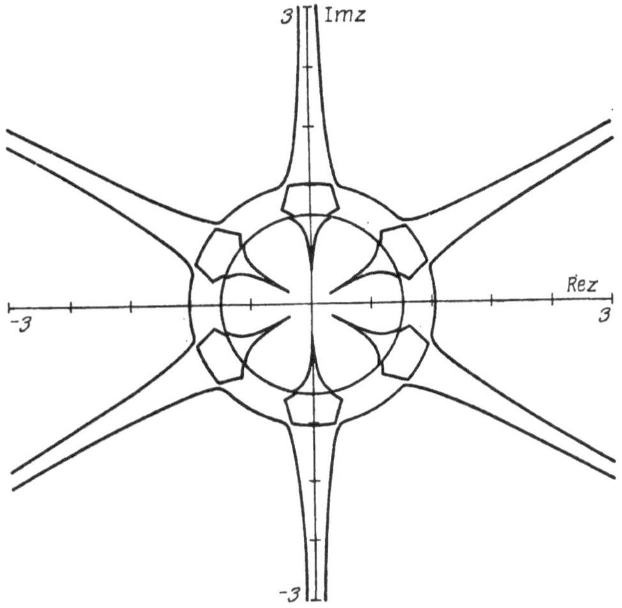

Bild 18. Der Einschluß für $T(z) = 1 - 3\varepsilon z^4/2 + z^6$, $|\varepsilon| = 1$ nach (6.9)

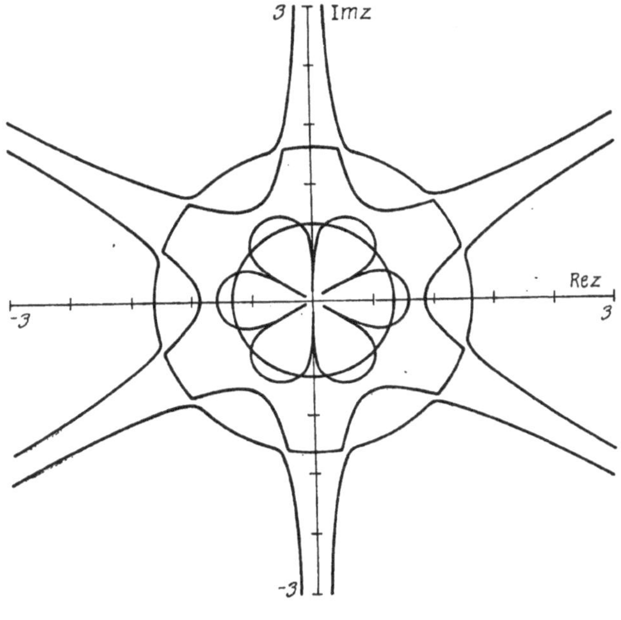

Bild 19. Der Einschluß für $T(z) = 1 - 5\varepsilon z^4/2 + z^6$, $|\varepsilon| = 1$ nach (6.9)

Literatur

1 Biernacki, M., Sur les équations algébriques contenant des paramètres arbitraires, Bull. Acad. Polonaise, Cl. Sci. Math., série A, 541–685 (1927).
2 Dieudonné, J., La théorie analytique des polynômes d'une variable (à coefficients quelconques), Gauthier-Villars, Paris 1938.
3 Marden, M., The Geometry of the Zeros of a Polynomial in a Complex Variable, Amer. Math. Soc. surveys III, New York 1949
4 Pólya, G., Geometrisches über die Verteilung der Nullstellen gewisser transzendenter Funktionen, S.-Ber. Bayer. Akad. Wiss., München 1920, 285–290.
5 Specht, W., Algebraische Gleichungen mit reellen oder komplexen Koeffizienten, Enz. d. math. Wiss. I1, Heft 3, II, B. G. Teubner, Stuttgart 1958.
6 Walsh, J. L., The Location of Critical Points of Analytic Functions and Harmonic Functions, Amer. Math. Soc. Coll. Publ. Vol. XXXIV New York 1950.
7 Melvin, W. R., Stability properties of functional difference equations, J. Math. Analysis Appl. 48 (1974), 749–763.
8 Avellar, C. E.; Hale, J. K., On the zeros of exponential polynomials J. Math. Analysis Appl. 73 (1980), 434–452.

Eingereicht: 17. 6. 1981

Anschrift: Dr. Günter Boese, Am Radenhölzl 8, D-8000 München 70, BRD

KLEINE MITTEILUNGEN

ZAMM 62, 561 – 563 (1982)

J. G. VERWER

A Note on a Runge-Kutta-Chebyshev Method

1. Introduction

This note deals with the numerical solution of the initial value problem for large systems of ordinary differential equations

$$y' = f(y) , \tag{1.1}$$

which possess the property that the eigenvalues of the Jacobian matrix $J(y) = \partial f(y)/\partial y$ are situated in a long narrow strip along the negative axis of the complex plane. Such systems frequently arise when discretizing the space variables of initial-boundary value problems for parabolic partial differential equations [7]. We shall focus our attention to these semi-discrete parabolic problems. For our discussion it is not necessary to define a particular class of problems or to specify the method of semi-discretization. Our only restriction is the location of the eigenvalues of $J(y)$. Further it is always assumed that the problem is sufficiently smooth.

Time integration methods for parabolic equations are distinguished into 3 classes: implicit methods, partly implicit or splitting methods, and explicit methods. Our method belongs to the latter class. In considering the use of explicit methods one must weigh up an advantage against a disadvantage. Their advantage is that they do not require the solution of large and complicated systems of nonlinear algebraic or transcendental equations (more dimensional problems). In fact, our methods are simple to use and can be easily applied to large problem classes. The sparsity structure of $J(y)$ is of no importance when using an explicit algorithm. For example, when setting aside the semi-discretization aspects, the same algorithm can be used if the space-dimension of the problem is changed or, when we are dealing with a system of parabolic equations, it is not too troublesome to vary the number of equations in this system. When using an implicit method one has to reckon with the structure of $J(y)$, because any efficient implicit solver exploits it.

Their disadvantage is of course the conditional stability. However, it is possible to reduce this disadvantage considerably by using so-called *stabilized Runge-Kutta methods* (cf. [2], section 2.7). Such a method uses a relatively large number of $f(y)$-evaluations per integration step, say m, the majority of which serves to enlarge the real stability boundary. As this boundary increases quadratically with m it certainly pays to employ such a stabilized method instead of a standard explicit one. Although stabilized explicit methods will often be less efficient than advanced implicit solvers in terms of computer time, this stabilization property, in combination with their simplicity and feasibility on large problem classes, gives us rise to support them.

In section 2 we discuss a class of 3-step Runge-Kutta methods which is similar to the class earlier reported in [11, 12]. These last methods, however, have two disadvantages. The first is that the integration parameters are not known in closed form. The second, and the most severe one, is that they are internally unstable. That is, within one single integration step they exhibit a severe accumulation of rounding errors, even when satisfying the condition of absolute stability. Because of the fact that this accumulation can easily influence the local accuracy it is desirable to develop internally stable methods. Recently, VAN DER HOUWEN and SOMMEIJER [3] reported stabilized one-step Runge-Kutta methods which are indeed internally stable for all values of m. They obtained internal stability after identifying each intermediate stability function with a Chebyshev polynomial via a stable, two-step Chebyshev recursion. Following this idea we developed a class of internally stable, 3-step Runge-Kutta methods. Our absolute stability boundaries are approximately three times larger than the boundaries reported by VAN DER HOUWEN and SOMMEIJER.

Section 3 deals with a linearization (cf. [10, 13]). By linearization we can replace $m - 1$ evaluations of $f(y)$ by $m - 1$ multiplications of a Jacobian matrix with intermediate vectors and, usually, one Jacobian matrix evaluation. This means that for many problems the linearized schemes will be less expensive.

Section 4 is devoted to some numerical experiments with a non-linear, 2-space dimensional parabolic problem. The aim of this section is twofold. In the first place we wish to illustrate the numerical performance of the stabilized formulas, and the linearizations, on a non-trivial problem. Secondly, we wish to give some insight in their efficiency when compared with an advanced implicit solver. We therefore also give results of the wellknown 4-th order backward differentiation formula, equipped with a Newton process and an iterative method whose convergence is accelerated by using coarse grid corrections [4]. As expected, for our 2-space dimensional example problem the explicit formulas are not competitive when comparing the accuracy against the computational work.

2. A three-step Runge-Kutta-Chebyshev method

In this section we briefly describe the *stabilized, explicit method* we deal with. As the construction of the method is highly technical, and rather lengthy, details are omitted (see [13]).

Let y_n denote the numerical approximation at $t = t_n$. Let $\tau = t_{n+1} - t_n$ denote a *stepsize*. Our *integration formula* then reads

$$\begin{aligned}
y_{n+1}^{(0)} &= \mu_0 y_n + (1 - \mu_0)\, y_{n-1} ,\\
y_{n+1}^{(1)} &= \mu_1 y_n + (1 - \mu_1)\, y_{n-1} + \tau(\tilde{\gamma}_1 f(y_n) + \tilde{\delta}_1 f(y_{n-1})) ,\\
y_{n+1}^{(j)} &= \mu_j y_{n+1}^{(j-1)} + (1 - \mu_j)\, y_{n+1}^{(j-2)} + \tau \tilde{\mu}_j f(y_{n+1}^{(j-1)}) , \qquad j = 2(1)\, m ,\\
y_{n+1} &= \alpha(\alpha_0 y_{n+1}^{(m)} + \alpha_1 y_n + \alpha_2 y_{n-1}) + (1 - \alpha)\, y_{n-2} , \qquad m \geqq 2 .
\end{aligned} \tag{2.1}$$

Formula (2.1) is readily recognized as a 3-step integration formula using m function evaluations per step. We confine ourselves to formulas of order of consistency $p \leqq 2$ and concentrate on stability aspects, i.e. the coefficients in (2.1) are chosen in such a way that the absolute stability regions contain a narrow strip along the negative axis in $\mathbb{C}$ of almost optimal length. The coefficients given below define such regions, where the absolute stability boundaries, say $\beta_p(m)$, are given by

$$\beta_p(m) \simeq (7.9 - 2.8p)\, m^2 , \qquad m \geqq 2 . \tag{2.2}$$

As the effective boundary $\beta_p(m)/m$, i.e. the boundary per evaluation, increases linearly with m, we wish to reckon with large values for m. Therefore we have to take care of internal stability or stability within one single step. Internal stability is a property of the recurrence equation for $y_{n+1}^{(j)}$, $j \geqq 2$. In our formulas the coefficients μ_j, $\tilde{\mu}_j$ are chosen in such a way that for the model equation $y' = \delta y$, $\delta < 0$, a stable 3-term Chebyshev recursion appears.

Both for order $p = 1$ and $p = 2$, the coefficients are given by

$$\begin{aligned}
&\mu_0 = \mu_1 = (a + b(1 - p_0))/2a , \qquad \tilde{\gamma}_1 = w_1 \mu_0 / w_0 ,\\
&\tilde{\delta}_1 = w_1(1 - \mu_0)/w_0 , \qquad \mu_j = 2 w_0 T_{j-1}(w_0)/T_j(w_0) ,\\
&\tilde{\mu}_j = 2 w_1 T_{j-1}(w_0)/T_j(w_0) , \qquad j = 2(1)\, m ,\\
&\alpha = 2/(2 - p_0) , \qquad \alpha_0 = 2a ,\\
&\alpha_1 = (1 - b)\,(1 - p_0) - a , \qquad \alpha_2 = 1 - 2a - \alpha_1 ,
\end{aligned} \tag{2.3}$$

where

$$w_0 = 1 + 0.05/m^2, \qquad w_1 = (\tfrac{1}{2} - \tfrac{1}{4} p_0)\, T_m(w_0)/aT'_m(w_0). \tag{2.4}$$

Here $T_j(x) = \cos[j \arccos x]$. The parameters a, b and p^0 determine the order of consistency of the scheme. For the first order formulas we have $a = 0.975$, $b = 0.2$, $p_0 = 124/229$, and for order $p = 2$, $a = 0.81$, $b = 0.2$ while p_0 satisfies ($p_0 \simeq -0.66$)

$$\left(\frac{b}{a} + \frac{\xi}{4}\right) p_0^2 - \left(\frac{3b}{a} + \xi\right) p_0 + \xi + \frac{2b}{a} - 4 = 0, \qquad \xi = \frac{T_m(w_0)\, T''_m(w_0)}{a T'^2_m(w_0)}. \tag{2.5}$$

When using these coefficients it appears that the error constants of (2.1) are almost independent of m. Finally we observe that (2.1) needs 7 arrays of memory places of the length of the vector y.

3. Linearized Runge-Kutta-Chebyshev formulas

Suppose we are given a semi-discrete parabolic equation (1.1) for which it is not too cumbersome to write down the Jacobian $J(y)$, and for which an $f(y)$-evaluation requires significantly more computational work than a multiplication of $J(y)$ with some vector. For such problems the costs of one integration step with (2.1), when performed with a large value of m, can be reduced considerably by replacing $f(y_{n+1}^{(j-1)})$ by the linearization (cf. [10, 13])

$$f(y_n) + J(y_{n+\eta})\,(y_{n+1}^{(j-1)} - y_n). \tag{3.1}$$

Substitution of this expression into (2.1) yields an integration formula using one $f(y)$-evaluation, $m-1$ matrix-vector multiplications, and, possibly, one $J(y)$-evaluation. In (3.1) the vector $y_{n+\eta}$ is assumed to be a first order approximation at some point $t = t_n + \eta\tau$. Because our formulas (2.1) are of order $p \leqq 2$, substitution of (3.1) does not reduce this order.

The parameter η has been introduced in order to indicate that we have some freedom in the choice of the evaluation point of J. In particular, it shows that when the Jacobian matrix is kept fixed during some finite number of integration steps, the order of consistency remains unchanged. In this case the error constants grow at each step, however, implying that the Jacobian matrix should be kept fixed only if the problem is weakly nonlinear. If $\eta = 0$, the error constants of the linearized formulas are close to the original ones. Usually, we therefore prefer to compute $J(y)$ at each step at the point $y = y_n$. The results presented in the next section support this strategy, i.e., for the problem in the example the linearization does not reduce the accuracy. Furthermore, as m will usually be large, the extra costs of one $J(y)$-evaluation per step can be justified.

4. Numerical experiments

Our parabolic example problem is given by (cf. [7], p. 201)

$$u_t = \left(\frac{\partial^2}{\partial x_1^2} + \frac{\partial^2}{\partial x_2^2}\right)(u^{\nu+1}), \qquad 0 < t \leqq 1, \quad 0 < x_1, x_2 < 1, \tag{4.1}$$

where we assume Dirichlet boundary conditions. As a solution we take $u(t, x_1, x_2) = a(bt + x_1 + x_2)^{1/\nu}$, where a and b are constant and $b = 2a^\nu(\nu+1)\,\nu^{-1}$, $\nu \neq 0$. By imposing a uniform grid on $\{(x_1, x_2) \mid 0 \leqq x_1, x_2 \leqq 1\}$, having $(h^{-1}-1)^2$ interior gridpoints, and by replacing the Laplacian by the standard 5-point difference operator, we obtain the *semi-discrete system*

$$\dot U_{ij} = h^{-2}\left[U_{ij-1}^{\nu+1} + U_{i-1j}^{\nu+1} - 4U_{ij}^{\nu+1} + U_{i+1j}^{\nu+1} + U_{ij+1}^{\nu+1}\right], \qquad i, j = 1(1)(h^{-1}-1), \tag{4.2}$$

where $U_{ij}(t) \simeq u(t, ih, jh)$ and where for $i, j \in \{1, h^{-1}-1\}$ at the appropriate places the exact boundary values have to be substituted. System (4.2) is thus of the form

$$\dot Y = g(t, Y), \tag{4.3}$$

$\partial g/\partial Y$ having the structure of the 5-point difference operator, i.e., the row corresponding to U_{ij} reads (zeros for boundary values)

$$\frac{\nu+1}{h^2}\left(U_{ij-1}^\nu \ldots U_{i-1j}^\nu \;\; -4U_{ij}^\nu \;\; U_{i+1j}^\nu \ldots U_{ij+1}^\nu\right). \tag{4.4}$$

If we write (4.3) in the autonomous form (1.1), the Jacobian matrix

$$J(y) = \begin{bmatrix} \dfrac{\partial g}{\partial Y} & \dfrac{\partial g}{\partial t} \\ 0 \ldots 0 & 0 \end{bmatrix}, \qquad y = [Y, t]^{\mathrm T}, \tag{4.5}$$

results. Note that, $\partial g/\partial Y$ is non-symmetric. In our experiments all explicit methods made use of the autonomous form, whereas the implicit solver was directly applied to (4.3).

The implicit solver we compare with has been proposed by VAN DER HOUWEN and DE VRIES [4]. They discuss an algorithm based on the 4-th order backward differentiation formula

$$Y_{n+1} - \frac{48}{25} Y_n + \frac{36}{25} Y_{n-1} - \frac{16}{25} Y_{n-2} + \frac{3}{25} Y_{n-3} = \frac{12}{25}\tau g(t_{n+1}, Y_{n+1}). \tag{4.6}$$

It would go too far to give a precise description of this algorithm. In short, by means of the Newton process the non-linear implicit equation (4.6) is replaced by a sequence of linear equations to which the incomplete LU-iteration method [6] is applied. This iterative method is then implemented in a two-level scheme, i.e., the convergence of this method is accelerated by using coarse grid corrections. We choose this particular implicit algorithm because the authors of [4] also used problem (4.1) as a test example. Apart from this, a comparison with the algorithm of MANTEUFFEL [5] gives strong evidence that their algorithm is very promising with respect to computational efficiency. A disadvantage is that it is rather complicated and that the storage requirement is considerable. VAN DER HOUWEN and DE VRIES discuss several variants of their algorithm. The variant we deal with is, in their notation, denoted by $(\mathrm{EIC_4I})^k$. Roughly speaking, this means that at each integration step with (4.6), k (prescribed) Newton iterations are performed. Each of these involves 1 evaluation of $g(t, Y)$ (E), 1 incomplete LU-iteration step on the h-grid (I), a coarse grid correction ($\mathrm{C_4}$) involving 4 incomplete LU-iteration steps on the $2h$-grid, and, finally, another incomplete LU-iteration on the h-grid (I). Further, each step with $(\mathrm{EIC_4I})^k$ involves one $\partial g/\partial Y$-evaluation.

A numerical comparison between method (2.1) and the related 1-step Runge-Kutta-Chebyshev method has already been reported in [9]. From this comparison we learned that normally the one-step method is more accurate than the 3-step one. Despite the fact that the absolute stability boundaries of the 3-step schemes are approximately 3 times larger than the corresponding 1-step boundaries, the 3-step schemes did not show a larger efficiency. Taking into consideration that a 1-step method is also easier to implement, the non-linearized 1-step Runge-Kutta-Chebyshev method should be preferred to the 3-step one. For the linearized methods, however, the situation is entirely different. Therefore, for the sake of completeness, we also insert results obtained with the second order explicit 1-step method, linearized and non-linearized (the first order 1-step method falls behind the second order one).

The condition of absolute stability for the explicit methods is, at each step, $\tau\sigma(J(y)) \leqq \beta_p(m)$, where σ denotes the spectral radius. Using (4.4) and the exact solution u of (4.1), we arrive at the estimate $\sigma(t) \simeq 4\nu b(bt + 2)\, h^{-2}$. We always choose the minimal m still satisfying $\tau\sigma(t_{n+1}) \leqq \beta_p(m)$. Hence, m varies per step. In practical situations it is of course desirable to have an automatic check on stability (cf. [12]).

For our experiments we selected $h = 1/20$ and $a = 0.8$, $b = 2$, $\nu = 4$. We thus have to deal with a rather non-linear problem possessing a smooth solution. Note that we also have relatively large values for $\sigma(t)$ which is of course not attractive for the explicit methods. The results of the experiments have been brought together into 2 accuracy-computational effort plots. The accuracy is measured by $sd = -{}^{10}\log$ (max. abs. error at $t = 1$). In Fig. a the computational effort is measured by fev = the total number of derivative evaluations and in

Fig. b, where results of the linearizations with $\eta = 0$ are given, the effort is measured by $mv =$ the total number of matrix-vector multiplications $+4\tau^{-1}$. Here $4\tau^{-1}$ should be considered as a correction term representing the number of derivative and

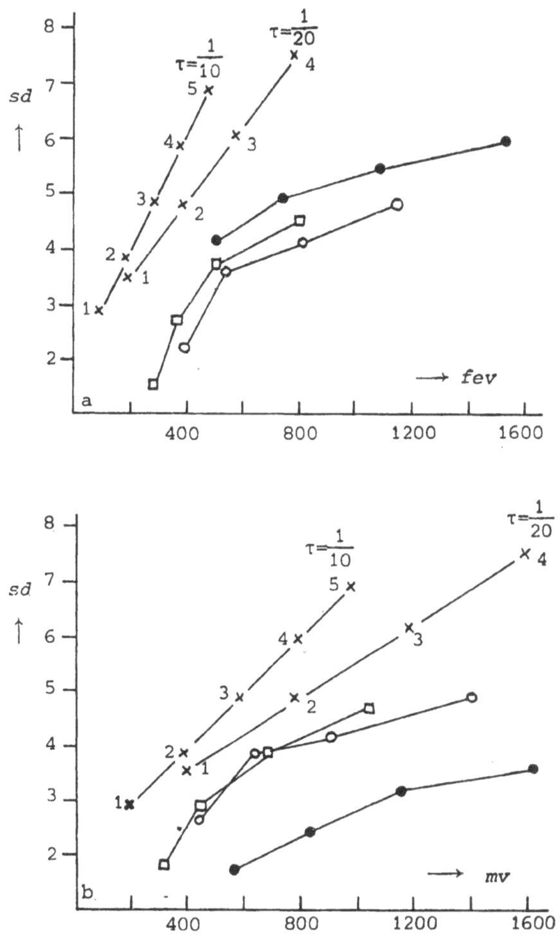

Fig. a, b. ●——● second order 1-step method; ○——○ second order 3-step method; □——□ first order 3-step method; ×——× algorithm $(EIC_4I)^k$, the integers refer to k. The results with the 1-step and 3-step method were obtained with $\tau = 1/5, 1/10, 1/20, 1/40$ and $\tau = 1/10, 1/20, 1/40, 1/80$, respectively

Jacobian evaluations. To be able to perform a comparison between the simple explicit methods and the complicated algorithm $(EIC_4I)^k$, the effort of this algorithm is also measured in the units derivative evaluation and matrix-vector multiplication. In Fig. a we accounted $10k$ derivative evaluations for 1 step with $(EIC_4I)^k$. This number was suggested in [4]. In Fig. b we thus accounted $20k$ matrix-vector multiplications. It shall be clear that there is some arbitrariness behind these choices.

Finally, the extra starting vectors were obtained from the exact solution of (4.1) which is also used for the error measuring. In the present situation this does not lead to misleading conclusions because in all integrations, with the exception of the one with $(EIC_4I)^4$ where $\tau = 1/20$, the time-integration error dominates the space-discretization error. The advantage of the extra starting vectors has been settled in the plots.

From Figures a, b we can conclude that all integrations have been performed successfully, that is, without numerical instabilities. In Figure a the explicit 1-step scheme is clearly more efficient than both explicit 3-step schemes. This conclusion does not apply to Figure b. On the contrary, after linearization the accuracy of the 1-step scheme has been strongly decreased, whereas the 3-step schemes maintain their accuracy. We noticed a similar behaviour on other example problems, that is, when using linearization the 3-step schemes turn out to be superior, whereas without linearization the 1-step scheme should be preferred. Also note that there is not much difference in the results of the two 3-step schemes. Evidently, this is caused by the larger stability boundary of the first order one.

Without doubt the implicit solver $(EIC_4I)^k$ takes advantage of the 4-th order of formula (4.6), even for small values of k. In both figures $(EIC_4I)^k$ stands out to be superior. Though there is some arbitrariness in our comparison it is expected that this conclusion applies to most scalar parabolic problems in 2 space dimensions. A serious impediment for using $(EIC_4I)^k$ could be its considerable storage requirement [4]. If one is tied to a computer system charging the need of much storage, the final costs of a run with $(EIC_4I)^k$ might be disappointing. In such situations the application of Runge-Kutta-Chebyshev methods could be considered.

To clarify this point we have listed the dependence of memory supply on the mesh width h of the three methods for a single, 2-space dimensional parabolic problem (see Table c). Here we considered the standard 5-point coupling on a uniform grid containing h^{-2} gridpoints approximately.

Table c
Dependence of memory supply on h for the standard 5-point coupling on a uniform grid. The bracketed quantities belong to the linearized schemes.

$(EIC_4I)^k$	the 3-step scheme (2.1)	the 1-step scheme from [3]
$24h^{-2}$	$7h^{-2}$ $(12h^{-2})$	$5h^{-2}$ $(10h^{-2})$

Finally, a remark on the explicit rational Runge-Kutta methods [14]. HAIRER [1] investigated the use of such methods for semi-discrete parabolic equations. He proved unconditional stability for linear homogeneous systems having a normal Jacobian matrix. However, a drawback of these rational methods seems to be their sensitiveness to non-normality of the Jacobian matrix. A numerical comparison [8] between an unconditionally stable second order method ([1], formula (5.1)) and the 1-step, second order Runge-Kutta-Chebyshev method confirmed this. For difficult problems the rational methods can not compete with the stabilized explicit ones.

Acknowledgement

The author wishes to acknowledge the many helpful remarks and programming assistence of Mr. B. P. SOMMEIJER.

References

1 HAIRER, E., Unconditionally stable explicit methods for parabolic equations, Numer. Math. **35**, 57–68 (1980).
2 HOUWEN, P. J. VAN DER, Construction of Integration Formulas for Initial Value Problems, Amsterdam, North Holland Publishing Company 1977.
3 –; SOMMEIJER, B. P., On the internal stability of explicit, m-stage Runge-Kutta methods for large values of m, ZAMM **60**, 479–485 (1980).
4 –; VRIES, H. B. DE, Preconditioning and coarse grid corrections in the solution of the initial value problem for nonlinear partial differential equations, SIAM J. Sci. Statist. Comput., to appear.
5 MANTEUFFEL, T. A., The Chebyshev iteration for non-symmetric linear systems, Numer. Math. **28**, 307–327 (1977).
6 MEIJERINGH, J. A.; VORST, H. A. VAN DER, An iterative solution method for linear systems of which the coefficient matrix is a symmetric M-matrix, Math. Comput. **31**, 148–162 (1977).
7 RICHTMYER, R. D.; MORTON, K. W., Difference Methods for Initial-Value Problems, New York Interscience 1969.
8 SOMMEIJER, B. P., Private communication.
9 –; VERWER, J. G., A performance evaluation of a class of Runge-Kutta-Chebyshev methods for solving semi-discrete parabolic differential equations, Report NW 91/80, Mathematical Centre, Amsterdam 1980.
10 –; HOUWEN, P. J. VAN DER, On the economization of stabilized Runge-Kutta methods with applications to parabolic initial value problems, ZAMM **61**, 105–114 (1981).
11 VERWER, J. G., A class of stabilized three-step Runge-Kutta methods for the numerical integration of parabolic equations, J. Computat. Appl. Math. **3**, 155–166 (1977).
12 –, An implementation of a class of stabilized, explicit methods for the time integration of parabolic equations, TOMS **6**, 188–205 (1980).
13 –, On a class of explicit three-step Runge-Kutta methods with extended real stability intervals, Report NW 77/79, Mathematical Centre, Amsterdam 1980.
14 WAMBECQ, A., Rational Runge-Kutta methods for solving systems of ordinary differential equation, Computing **20**, 333–342 (1978).

Eingereicht am 4. 8. 1981, korrigierte Fassung am 29. 10. 1981

Anschrift: Dr. J. G. VERWER, Stichting Mathematisch Centrum, Kruislaan 413, 1098 SJ Amsterdam, The Netherlands

ZAMM 62, 564–565 (1982)

P. CARRAGHER / L. J. CRANE

Heat Transfer on a Continuous Stretching Sheet

1. Introduction

Many industrial processes involve the cooling of continuous strips by drawing through a quiescent fluid; the drawing of sheet glass is an example of such a process. The properties of the final product depend to a large extent on the rate at which the material is cooled. Now the heat tranfer from a continuous sheet which moves at constant speed and is maintained at constant temperature has been theoretically discussed by TSOU, SPARROW and GOLDSTEIN [1]; these authors give in addition details of experiments on the laminar velocity field and on the velocity and temperature fields when the flow is turbulent. The experimental evidence is shown to be in good agreement with theory. In practice however these results are not directly applicable since neither the speed of the sheet, nor its temperature is constant along its length.

This paper is a theoretical investigation into the heat transfer from a stretching sheet which issues from a slit into a medium at rest; the velocity of the sheet is proportional to the distance from the slit; the temperature difference between the sheet and its surroundings is proportional to a power of the distance from the slit. Rates of heat transfer are obtained for moderate and large Prandtl numbers.

2. Equations of motion

Rectangular Cartesian coordinates (x, y) are taken with origin at the slit; x is measured along the sheet in the direction of motion; (u, v) are the corresponding velocity components. When the velocity of the sheet is $U_p = \alpha x$, where α is a constant, the velocity components of the flow in the boundary layer take the simple form:

$$u = \alpha x \exp(-\beta y) \tag{1}$$

$$v = -\alpha \frac{[1 - \exp(-\beta y)]}{\beta}, \tag{2}$$

where $\beta = \sqrt{\alpha/\nu}$, and ν is the kinematic viscosity; see CRANE [2].
The boundary layer equation governing the flow of heat is:

$$u \frac{\partial T}{\partial x} + v \frac{\partial T}{\partial y} = \frac{\nu}{P} \frac{\partial^2 T}{\partial y^2} \tag{3}$$

where T is the temperature of the fluid and P is the Prandtl number. The boundary conditions satisfied by T are:

at the edge of the boundary layer: $T = \text{constant} = T_\infty$, on the surface of the sheet:

$$T = T_\infty + A T_\infty x^n \equiv T_S \tag{4}$$

where A is a constant.

3. Temperature distribution at moderate values of P

It is convenient to introduce the boundary layer variable η defined by

$$\eta = \beta y \tag{5}$$

and to assume a similarity form for the temperature namely:

$$\frac{T - T_\infty}{T_\infty} = A x^n g(\eta). \tag{6}$$

Then equation (3) reduces to

$$P\left[n \exp(-\eta)\, g - (1 - \exp(-\eta)) \frac{dg}{d\eta}\right] = \frac{d^2 g}{d\eta^2} \tag{7}$$

with boundary conditions:

$$g(0) = 1 \quad \text{and} \quad g(\infty) = 0. \tag{8}$$

Equation (7) may be simplified by the change of variable

$$t = e^{-\eta}; \tag{9}$$

thus:

$$t \frac{d^2 g}{dt^2} + \frac{dg}{dt}[1 + P(t - 1)] - nPg = 0 \tag{10}$$

with boundary conditions:

$$g = 1 \quad \text{at} \quad t = 1 \quad \text{and} \quad g = 0 \quad \text{at} \quad t = 0. \tag{11}$$

The solution of (10) subject to (11) is

$$g = a t^P F(P - n, P + 1; -Pt) \tag{12}$$

where

$$a = [F(P - n, P + 1; -p)]^{-1} \tag{13}$$

and where F denotes the confluent hypergeometric functions, see [3].

4. Special cases

The solution of (12) reduces to a simple form in the following cases: $P = 1$, $n = 0$:

$$g = \frac{e}{e - 1}[1 - \exp(-e^{-\eta})], \tag{14}$$

$P = n$:

$$g = e^{-P\eta}, \tag{15}$$

$n = -1$, all P:

$$g = t^P e^{P - Pt}. \tag{16}$$

5. Temperature distribution for large values of P

When P is less than unity the temperature profile is readily calculated from the series (12). However when P is very much larger than unity an alternative form of solution is needed. Since the thickness of the temperature layer is proportional to $P^{-1/2}$ when P is large it is convenient to introduce the change of variable

$$\tau = \eta \sqrt{P}. \tag{17}$$

Then equation (7) becomes:

$$\frac{d^2 g}{d\tau^2} + \left[1 - \exp\left(\frac{-\tau}{\sqrt{P}}\right)\right] \sqrt{P} \frac{dg}{d\tau} - n \exp\left(\frac{-\tau}{\sqrt{P}}\right) g = 0. \tag{18}$$

Expanding the coefficients of g, $\frac{dg}{d\eta}$ in series form gives:

$$\frac{d^2 g}{d\tau^2} + \left[\tau - \frac{\tau^2}{2\sqrt{P}} + \dots\right] \frac{dg}{d\tau} - n\left[1 - \frac{\tau}{\sqrt{P}} + \dots\right] g = 0. \tag{19}$$

The solution of (19) may be found in the form:

$$g = g_1(\tau) + \frac{1}{\sqrt{P}} g_2(\tau) + \frac{1}{P} g_3(\tau) + \dots. \tag{20}$$

It is readily shown that

$$g_1 = F\left(\frac{-n}{2}, \frac{1}{2}; \frac{-\tau^2}{2}\right) + a_1 \tau F\left(\frac{-n}{2} + \frac{1}{2}, \frac{3}{2}, \frac{-\tau^2}{2}\right) \tag{21}$$

where

$$a_1 = -\sqrt{2}\,\Gamma\left(\frac{n}{2} + 1\right) \Big/ \Gamma\left(\frac{n}{2} + \frac{1}{2}\right), \tag{22}$$

(where Γ denotes the gamma function, see [3]), satisfies the conditions $g_1(0) = 1$, $g_1(\infty) = 0$,; the function g_2 which satisfies the conditions $g_2(0) = g_2(\infty) = 0$ is

$$g_2 = -\frac{1}{6}\left[(\tau^2 + 4n + 2)\frac{dg_1}{d\tau} + \{2n\tau - (4n + 2)\,a_1\}\,g_1\right]. \quad (23)$$

6. The Nusselt number Nu_x

The heat transfer from one side of the sheet may be expressed in terms of the Nusselt number defined by

$$\begin{aligned} Nu_x &= -\left(\frac{\partial T}{\partial y}\right)_0 \Big/ (T_S - T_\infty) \\ &= -\sqrt{\frac{Up}{\nu x}}\left[\frac{1}{g}\frac{dg}{d\eta}\right]_{\eta=0}. \end{aligned} \quad (24)$$

Some special cases include:

$$P = 1, n = 0:\ Nu_x \Big/ \sqrt{\frac{U_P}{\nu x}} = \frac{1}{e - 1}$$

$$P = n:\ Nu_x \Big/ \sqrt{\frac{U_P}{\nu x}} = P\,.$$

$$n = -1, \text{ all } P:\ Nu_x \Big/ \sqrt{\frac{U_P}{\nu x}} = 0\,.$$

In general:

$$Nu_x \Big/ \sqrt{\frac{U_P}{\nu x}} = P\left[1 - \frac{P - n}{P + 1}\frac{F(P - n + 1, P + 2; -P)}{F(P - n, P + 1; -P)}\right], \quad (25)$$

a convenient formula for $P < 1$.

When $P \gg 1$, the solution (20) gives:

$$\begin{aligned} \frac{Nu_x}{\sqrt{\frac{U_P}{\nu x}}} &= -\sqrt{P}\left[\frac{dg_1}{d\tau} + \frac{1}{\sqrt{P}}\frac{dg_2}{d\tau} + \dots\right]_{\tau=0} \\ &= -\sqrt{P}\left[a_1 + \frac{1}{6\sqrt{P}}\{(4n+2)\,a_1^2 - (4n^2+4n)\}\right], \end{aligned} \quad (26)$$

where a_1 is defined by (22).

7. Discussion

When P is less than unity a few terms of the series (25) give the heat transfer to good accuracy. For large Prandtl numbers (26) is more convenient. Since (25) and (26) agree to within one or two percent when $P = 1$ it follows that (26) may be used when $P \geqq 1$.

Table 1 contrasts typical cases of heat transfer in stretching and non-stretching sheets respectively.

Table 1: Values of $Nu_x \Big/ \sqrt{\frac{U_P}{\nu x}}$; the non-stretching cases are taken from reference [1].

	Stretching	Non-stretching
$n = 0$, $P = 1$	.58	.44
$n = 0$, $P = .7$	.46	.35
$n = -1/2$, $P = .7$	.26	0

References

1 TSOU, F. K.; SPARROW, E. M.; GOLDSTEIN, R. J., Flow and Heat Transfer in the Boundary Layer on a Continuous Moving Surface, Int. J. Heat Mass Transfer 10 (1967) p. 219.

2 CRANE, L. J., Flow Past a Stretching Plate, Journal of Applied Mathematics and Physics (ZAMP) 21 (1970) pp. 645–647.

3 JAHNKE, E.; EMDE, F., Tables of Functions, Dover, New York (1943).

Eingereicht am 24. 6. 1981, in revidierter Fassung am 14. 12. 1981

Anschrift: Dr. LAWRENCE J. CRANE, University of Dublin 39 Trinity College, Dublin 2, Ireland

ZAMM 62, 565 – 568 (1982)

T. C. CHIAM

Micropolar Fluid Flow over a Stretching Sheet

This article presents a numerical solution for the two-dimensional flow of incompressible micropolar fluids caused by a porous wall which stretches parallel to itself with a velocity proportional to the distance from the origin. The heat transfer from such a wall is also briefly discussed.

1. Introduction

The theory of micropolar fluids was first introduced by ERINGEN [1] to model fluids with microstructures which cannot be adequately described by the classical Stokesian theory. In such fluids the microelements not only possess translational motion but can also undergo rotational motion. The interaction of the macro-velocity field v and the microrotation field σ is described through new material constants in addition to those for a classical Newtonian fluid. Some examples of such fluids are animal blood, liquid crystals and certain polymeric fluids.

Since the publication of ERINGEN's work [1], many authors have investigated various flow problems (see, for example [2] – [8]).

In a recent paper, CHAKRABARTI and GUPTA [9] studied the flow of an electrically conducting incompressible fluid the motion of which is caused by a porous plane wall stretching parallel to itself with velocity proportional to the distance from the origin in the presence of a transverse uniform magnetic field. The authors obtained a very simple exact solution.

The aim of the present work is to extend the nonconducting case of [9] to that of a micropolar fluid. As it is unlikely that analytical solutions to the resulting nonlinear, coupled boundary value problem can be obtained, a numerical solution is presented. The heat conduction case is also briefly discussed.

2. Formulation of the Problem

Consider a porous plane wall situated at $y = 0$ with a micropolar fluid occupying the region $y > 0$. Two equal and opposite forces are introduced along the x-axis and the wall is stretched so that its velocity is proportional to x. We suppose that the motion of the fluid is caused solely by this stretching motion and that the fluid is continuously sucked through the wall with constant velocity v_0.

For steady two-dimensional motion of micropolar fluids with constant fluid properties, the basic equations of motion are [7]

$$\varrho\left(u\frac{\partial u}{\partial x} + v\frac{\partial u}{\partial y}\right) = -\frac{\partial p}{\partial x} + (\mu + \varkappa)\left(\frac{\partial^2 u}{\partial y^2} + \frac{\partial^2 u}{\partial x^2}\right) + \varkappa\frac{\partial \sigma}{\partial y} \quad (2.1)$$

$$\varrho\left(u\frac{\partial v}{\partial x} + v\frac{\partial v}{\partial y}\right) = -\frac{\partial p}{\partial y} + (\mu + \varkappa)\left(\frac{\partial^2 v}{\partial x^2} + \frac{\partial^2 v}{\partial y^2}\right) - \varkappa\frac{\partial \sigma}{\partial x} \quad (2.2)$$

$$\varrho j\left(u\frac{\partial \sigma}{\partial x} + v\frac{\partial \sigma}{\partial y}\right) = \gamma\left(\frac{\partial^2 \sigma}{\partial x^2} + \frac{\partial^2 \sigma}{\partial y^2}\right) - 2\varkappa\sigma + \varkappa\left(\frac{\partial v}{\partial x} - \frac{\partial u}{\partial y}\right) \quad (2.3)$$

$$\frac{\partial u}{\partial x} + \frac{\partial v}{\partial y} = 0 \quad (2.4)$$

where the velocity vector is $\boldsymbol{v} = [u(x, y), v(x, y), 0]$ and the microrotation vector is $\boldsymbol{\sigma} = [0, 0, \sigma(x, y)]$; ϱ is the density; p is the pressure; μ is the viscosity, $\varkappa$ the microrotational coupling coefficient, γ a microrotational diffusivity and j the square of a length typical of the microstructure with $\gamma, \varkappa, \mu \geq 0$. When these constants are zero, equations (2.1) to (2.3) reduce to the usual Navier-Stokes equations.

In the problem considered here, these equations simplify to the following

$$u \frac{\partial u}{\partial x} + v \frac{\partial u}{\partial y} = \nu \frac{\partial^2 u}{\partial y^2} + K \frac{\partial \sigma}{\partial y} \tag{2.5}$$

$$J\left(u \frac{\partial \sigma}{\partial x} + v \frac{\partial \sigma}{\partial y}\right) = G \frac{\partial^2 \sigma}{\partial y^2} - 2\sigma - \frac{\partial u}{\partial y} \tag{2.6}$$

$$\frac{\partial u}{\partial x} + \frac{\partial v}{\partial y} = 0 \tag{2.7}$$

where $\nu = \frac{\mu + \varkappa}{\varrho}$, $K = \frac{\varkappa}{\varrho}$, $G = \frac{\gamma}{\varkappa}$, $J = \frac{j\varrho}{\varkappa}$

subject to the boundary conditions

$$y = 0: u = Cx,\ v = -v_0,\ \sigma = 0; \tag{2.8}$$

$$y \to \infty: u \to 0,\ \sigma \to 0 .$$

As in [9], we let $\eta = (C/\nu)^{1/2} y$ and

$$u = Cxf'(\eta),\ v = -(\nu C)^{1/2} f(\eta),\ \sigma = C^{3/2} \nu^{-1/2} x g(\eta) . \tag{2.9}$$

The continuity equation (2.7) is then identically satisfied. Substitution into Eq. (2.5) gives

$$(f')^2 - ff'' = f''' + K_1 g' \tag{2.10}$$

where $K_1 = K/\nu = \varkappa(/\varrho\nu) = \frac{\varkappa}{\mu + \varkappa}$.

Eq. (2.6) gives

$$K_3(f'g - fg') = K_2 g'' - 2g - f'' \tag{2.11}$$

where $K_2 = \frac{GC}{\nu}$ and $K_3 \doteq JC$.

The appropriate boundary conditions for Eq. (2.10) and (2.11) are

$$f(0) = -\frac{v_0}{(\nu C)^{1/2}} = R,\ f'(0) = 1,\ f'(\infty) = 0 , \tag{2.12}$$

$$g(0) = 0, \quad g(\infty) = 0 \tag{2.13}$$

where R is a suction parameter.

3. Results and Discussion

Equations (2.10) and (2.11) subject to the conditions (2.12) and (2.13) form a nonlinear boundary value problem which is difficult to solve analytically. The problem was solved numerically by using the fourth order Runge-Kutta method with Newton iteration to search for the values of $f''(0)$ and $g'(0)$.

There are four parameters in the problem, viz. K_1, K_2, K_3 and R. We note that K_1 may only vary in the range $0 \leq K_1 \leq 1$; K_2 and K_3 are nonnegative and only small values of K_3 are significant (see [7]). In Table 1 are shown the values of $f''(0)$ and $g'(0)$ for $K_1 = 0.01$ and for K_2 and R taking values 1, 2, 3 and 4. Table 2 shows the corresponding values of $f''(0)$ and $g'(0)$ when $K_1 = 1.0$. These two tables thus exhibit the behaviour of $f''(0)$ and $g'(0)$ corresponding to K_1 small and K_1 maximum. In Table 1, only one column for $f''(0)$ is shown because the values for $f''(0)$ when $K_3 = 0$ and $K_3 = 0.05$ are virtually identical up to the fifth decimal place. Similar behaviour is also observed in Table 2. The effects of K_2 and R on $g'(0)$ are both significant. Table 3 gives the corresponding values of $f''(0)$ and $g'(0)$ for a Newtonian fluid for comparison.

The effect of suction on the velocity profile $f'(\eta)$ is shown in Figs. 1 and 2. It is seen that the greater the value of R, the faster the decay of $f'(\eta)$ to zero. A comparison of Figs. 1 and 2 also shows that when $K_1 = 1.0$, f' approaches zero faster than for $K_1 = 0.01$.

Figure 3 gives graphs of the microrotation function $g(\eta)$ for $K_1 = 0.01$, $K_2 = 1.0$, $K_3 = 0.05$ and for $R = 0$, 2 and 4 respectively.

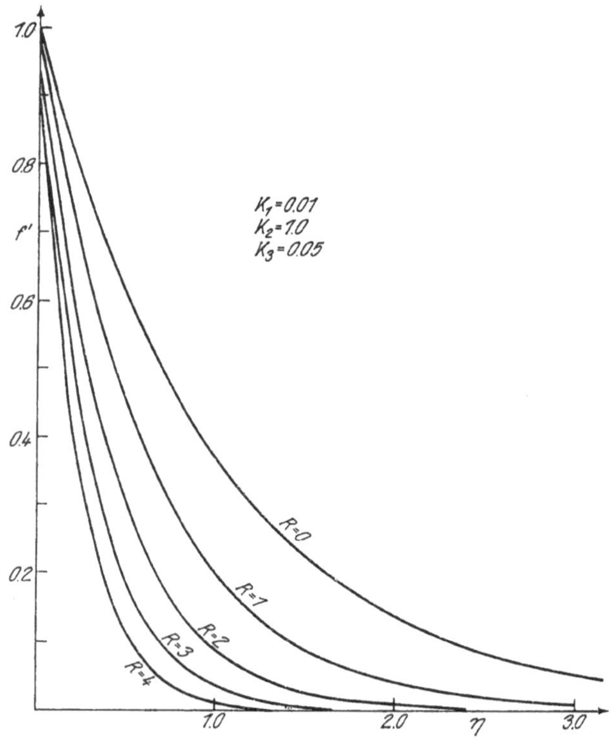

Fig. 1. Plots of f' vs. η for $R = 0, 1, 2, 3, 4$, $K_1 = 0.01$, $K_2 = 1.0$, $K_3 = 0.05$

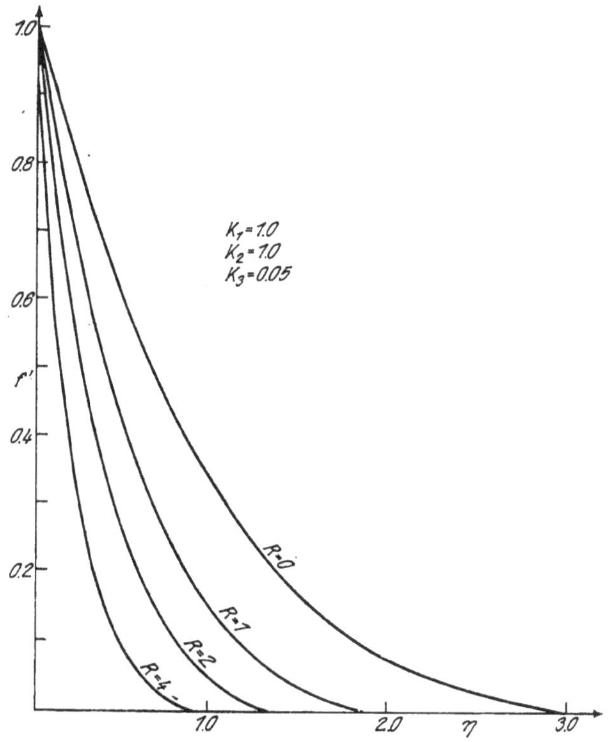

Fig. 2. Plots of f' vs. η for $R = 0, 1, 2, 4$, $K_1 = 1.0$, $K_2 = 1.0$, $K_3 = 0.05$

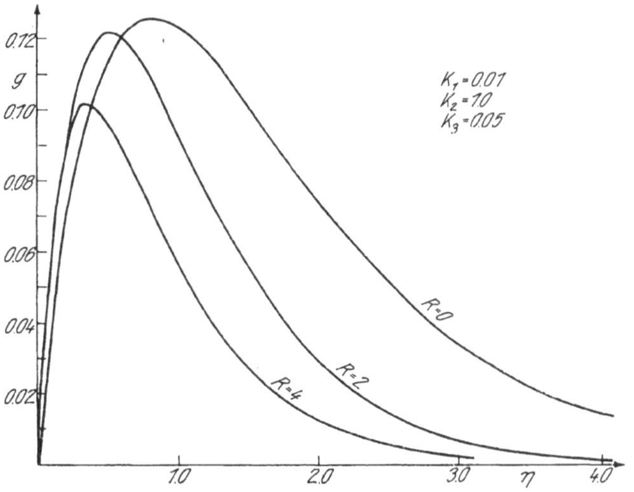

Fig. 3. Plots of microrotation function g vs. η for $R = 0, 2, 4$, $K_1 = 0.01$, $K_2 = 1.0$, $K_3 = 0.05$

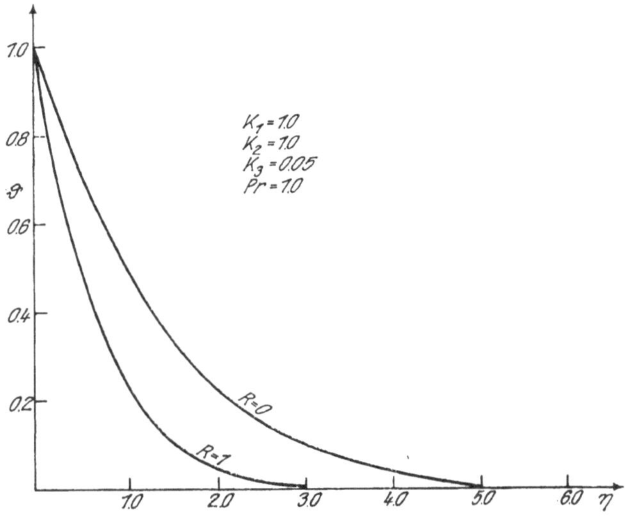

Fig. 4. Plots of dimensionless temperature θ vs. η for $R = 0, 1$, $K_1 = 1.0$, $K_2 = 1.0$, $K_3 = 0.05$ and Prandtl number = 1.0

4. Heat Transfer

The theory of heat conducting micropolar fluid has been presented by KAZAKIA and ARIMAN [10]. The heat equation derived is complicated. However, it is possible to simplify it for two-dimensional boundary layer flow [11] to the form

$$u \frac{\partial T}{\partial x} + v \frac{\partial T}{\partial y} = \varkappa \frac{\partial^2 T}{\partial y^2} + \alpha \left(\frac{\partial T}{\partial x} \frac{\partial g}{\partial y} - \frac{\partial T}{\partial y} \frac{\partial g}{\partial x} \right) \tag{4.1}$$

where the second term on the right represents interaction between the temperature field and the microrotation field.

Here, since T and g do not depend on x, Eq. (4.1) simplifies to the usual heat conduction equation. By introducing a non-dimensional temperature

$$\theta(\eta) = (T - T_\infty)/(T_w - T_\infty)$$

where $T = T_w$ at $y = 0$

and $T \to T_\infty$ as $y \to \infty$

the heat equation becomes

$$\theta'' = -(Pr) f \theta' \tag{4.2}$$

where Pr is the Prandt l number. The appropriate boundary conditions are then

$$\theta(0) = 1, \ \theta(\infty) = 0 . \tag{4.3}$$

Equation (4.2) satisfying Eq. (4.3) is a linear boundary value problem which is easily solved by superposition of solutions. As this problem is independent of g and as f is not sensitive to K_2 and K_3, we present in Table 4 values of $\theta'(0)$ only for $K_2 = 1.0$ and $K_3 = 0.05$ and $Pr = 1.0$. It is seen that the effect of K_1 on $\theta'(0)$ is slight but changes in R cause large changes in $\theta'(0)$. In Fig. 4 we show the plot of θ vs η for two values of R when $K_1 = 1.0$, $K_2 = 1.0$, $K_3 = 0.05$ and $Pr = 1$.

Table 1. Surface shear stress and microrotation gradient for $K_1 = 0.01$

K_1	K_2	R	$f''(0)$ $K_3 = 0, 0.05$	$g'(0)$ $K_3 = 0$	$g'(0)$ $K_3 = 0.05$
0.01	1.0	0.0	−0.99951	0.41445	0.41333
	2.0		−0.99971	0.25015	0.24963
	3.0		−0.99980	0.18360	0.18328
	4.0		−0.99986	0.14651	0.14628
0.01	1.0	1.0	−1.61765	0.53382	0.53719
	2.0		−1.61775	0.30910	0.31015
	3.0		−1.61782	0.22158	0.22210
	4.0		−1.81786	0.17400	0.17430
0.01	1.0	2.0	−2.41399	0.63078	0.63814
	2.0		−2.41407	0.35362	0.35589
	3.0		−2.41410	0.24912	0.25024
	4.0		−2.41413	0.19338	0.19405
0.01	1.0	3.0	−3.30265	0.70033	0.71068
	2.0		−3.30270	0.38384	0.38693
	3.0		−3.30272	0.26729	0.26878
	4.0		−3.30273	0.20592	0.20681
0.01	1.0	4.0	−4.23600	0.74982	0.76232
	2.0		−4.23603	0.40454	0.40818
	3.0		−4.23604	0.27948	0.28122
	4.0		−4.23604	0.21424	0.21526

Table 2. Surface shear stress and microrotation gradient for $K_1 = 1.0$

K_1	K_2	R	$f''(0)$ $K_3 = 0$	$f''(0)$ $K_3 = 0.05$	$g'(0)$ $K_3 = 0$	$g'(0)$ $K_3 = 0.05$
1.0	1.0	0.0	−0.93303	−0.93343	0.42863	0.42737
	2.0		−0.95333	−0.95354	0.25701	0.25645
	3.0		−0.96297	−0.96311	0.18781	0.18747
	4.0		−0.96886	−0.96896	0.14939	0.14916
1.0	1.0	1.0	−1.57745	−1.57752	0.55549	0.55890
	2.0		−1.59118	−1.59122	0.31771	0.31875
	3.0		−1.59721	−1.59725	0.22637	0.22688
	4.0		−1.60072	−1.60076	0.17708	0.17739
1.0	1.0	2.0	−2.39351	−2.39350	0.64985	0.65718
	2.0		−2.40110	−2.40110	0.36037	0.36260
	3.0		−2.40428	−2.40430	0.25267	0.25376
	4.0		−2.40607	−2.40609	0.19559	0.19624
1.0	1.0	3.0	−3.29182	−3.29180	0.71506	0.72523
	2.0		−3.29608	−3.29608	0.38870	0.39172
	3.0		−3.29779	−3.29780	0.26974	0.27120
	4.0		−3.29873	−3.29875	0.20742	0.20829
1.0	1.0	4.0	−4.22979	−4.22978	0.76105	0.77326
	2.0		−4.23234	−4.23234	0.40807	0.41162
	3.0		−4.23332	−4.23333	0.28123	0.28292
	4.0		−4.23385	−4.23387	0.21529	0.21629

Table 3. Surface shear stress and microrotation gradient for $K_1 = 0$ i. e. Newtonian fluid.

R	$f''(0)$	$g'(0)$
0.0	−1.00002	0.41323
1.0	−1.61808	0.53363
2.0	−2.41421	0.63060
3.0	−3.30278	0.70019
4.0	−4.23607	0.74971

Table 4. Values of $\theta'(0)$ when $K_2 = 1.0$, $K_3 = 0.05$, Pr = 1.0.

R	$\theta'(0)$ with $K_1 = 0.01$	$\theta'(0)$ with $K_1 = 1.0$
0.0	−0.58226	−0.56339
1.0	−1.32804	−1.31587
2.0	−2.21203	−2.20600
3.0	−3.13523	−3.14924
4.0	−4.11739	−4.11564

References

1 Eringen, A. C., Theory of micropolar fluids, J. Math. Mech. 16, 1–18 (1966).
2 Willson, A. J., The flow of a micropolar liquid layer down an inclined plane, Proc. Camb. Phil. Soc. 64, 513–526 (1968).
3 Willson, A. J., Basic flows of a micropolar liquid, Appl. Sci. Res. 20, 338–355 (1969).
4 Willson, A. J., Boundary layers in micropolar liquids, Proc. Camb. Phil. Soc. 67, 469–476 (1970).
5 Ariman, T.; Cakmak, A. S., Some basic viscous flows in micropolar fluids, Rheol. Acta 7, 236–242 (1968).
6 Gupta, P. S.; Gupta, A. S., Asymptotic suction problem in the flow of micropolar liquids, Acta Mechanica 15, 141–149 (1972).
7 Kümmerer, H., Similar laminar boundary layers in incompressible micropolar fluids, Rheol. Acta 16, 261–265 (1977).
8 Sastry, V. U. K.; Rao, V. R. M., Micropolar fluid flow due to an oscillating plane subject to rotation, Acta Mechanica, 33, 45–53 (1979).
9 Chakrabarti, A.; Gupta, A. S., Hydromagnetic flow and heat transfer over a stretching sheet, Quart. Appl. Math. 37, 73–78 (1979).
10 Kazakia, Y.; Ariman, T., Heat-conducting micropolar fluids, Rheol. Acta 10, 319–325 (1971).
11 Unsworth, K.; Chiam, T. C., Heat transfer from non-isothermal surfaces in the stagnation-point flow of a micropolar fluid, Rheol. Acta 19, 356–364 (1980).

Eingereicht am 16. 11. 1981

Anschrift: Dr. T. C. Chiam, School of Mathematical Sciences, University of Science, Penang, Malaysia

BUCHBESPRECHUNGEN

Schwalbe, K.-H., Bruchmechanik metallischer Werkstoffe. München-Wien, Carl Hanser Verlag 1980. 847 S., 500 Abb., DM 148,–. ISBN 3-446-12983-9

Das von K.-H. Schwalbe vorgelegte Buch ist eine sehr schöne Darstellung der wichtigsten Grundlagen und Anwendungsmöglichkeiten metallischer Werkstoffe. Es ist sicher schwierig, einen zusammenfassenden Überblick über dieses umfangreiche Gebiet geben zu wollen. Jedoch kann man dem Autor bescheinigen, daß er es gut verstanden hat, die komplizierten Bruch- und Rißvorgänge in Festkörpern auch in verständlicher Form darzubieten. Die Beschränkung auf das stabile Rißwachstum ist auch aus diesem Grunde sicher richtig. Hervorzuheben ist die gelungene Darstellung wichtiger Verbindungen zwischen makroskopisch-phänomenologischer Bruchmechanik und den Mikrovorgängen des Bruchs aus werkstoffphysikalischer Sicht. Die übersichtliche Darbietung des behandelten Stoffs, die durch zahlreiche aussagekräftige Abbildungen und Prinzipskizzen ausgezeichnet illustriert ist, kann man als gelungen bezeichnen. Ganz besonders hervorzuheben ist das ausführliche Quellenverzeichnis von nahezu 900 Literaturstellen, die auch dem Nichtspezialisten den Zugang zu den Originalarbeiten sehr erleichtern. Das Buch dürfte im deutschsprachigen Raum als eines der besten seiner Art gelten. Das Werk sollte insbesondere durch die vorwiegend mit Blickrichtung technischer Anwendungen orientierten Leser sehr gut aufgenommen werden. Aus der Sicht des stärker der Grundlagenforschung zur Problematik Bruch und Riß verpflichteten Forschers gibt es auch einige kritische Bemerkungen in dem Sinne, daß einige Sachverhalte zum Zwecke der Übersichtlichkeit der Darstellung stark vereinfacht dargestellt sind und bestimmte neuere Entwicklungen der letzten Zeit noch nicht berücksichtigt wurden. Diese Bemerkungen mindern den großen Wert des vorgelegten Werkes auch für den Spezialisten der Bruchforschung nicht. Die Verbindung zwischen den allgemeinen Grundlagen der linearelastischen und der Fließbruchmechanik und den speziellen Fragen, z. B. des Rißfortschritts, zu den Berechnungsmethoden angerissener Bauteile, z. B. Rissen in Schweißverbindungen usw., bis hin zu den experimentellen Meßmethoden der Ermittlung bruchmechanischer Kenngrößen, ist sehr gut gelungen. Das dürfte auch für den auf praktische Dimensionierungsaufgaben orientierten Berechnungsingenieur von Vorteil sein. Das Buch ist sehr gut für einen Einsatz in Forschung und Lehre geeignet. Es bleibt nur zu wünschen, daß Schwalbes „Bruchmechanik metallischer Werkstoffe" einen großen Leserkreis finden möge.

Berlin K. Hennig

Nemat-Nasser, S. (ed.), Three-Dimensional Constitutive Relations and Ductile Fracture. Proceedings, Dourdan, France 1980. Amsterdam-New York-Oxford, North-Holland Publ. Co. 1981. XV, 439 S., US $ 95.00. Dfl. 195.00. ISBN 0-444-86108-4

Die in der letzten Zeit von Nemat-Nasser herausgegebenen Bücher zur Festkörpermechanik zählen sämtlich zur international anerkannten Fachliteratur. Der vorliegende Band zu den „dreidimensionalen konstitutiven Beziehungen (Stoffgesetzen) und zum duktilen Bruch" bildet hier keine Ausnahme. Das Werk enthält die auf dem IUTAM Symposium zur gleichen Thematik 1980 in DOURDAN, Frankreich, gehaltenen eingeladenen Vorträge international führender Wissenschaftler der Mechanik der Bruchvorgänge. Die Arbeiten lassen sich in drei Themenbereiche einteilen:

1. Untersuchungen zur Entstehung und Ausbreiten von Rissen in duktilen Werkstoffen
2. Stoffgesetze für das inelastische Materialverhalten im rißspitzennahen Gebiet
3. makroskopisch-phänomenologische Schlußfolgerungen, die sich aus 1. und 2. ergeben.

Neben der Darlegung neuer und der kritischen Wertung bekannter theoretischer Konzepte der Bruch- und Rißmechanismen sind auch umfangreiche experimentelle Untersuchungen dargestellt und an Hand guter Abbildungen illustriert. Von der Seite der „Werkstoffmechanik" her ist der Standpunkt der Einheit von klassischer Bruchmechanik und Mechanik der Mikrovorgänge (Mikromechanik) in vielen Beiträgen unverkennbar. Solche Namen wie Argon, Mc Clintok und Nemat-Nasser oder Bilby sind in diesem Zusammenhang stellvertretend zu nennen. Die große Zahl kompetenter Bruch- und Rißforscher macht es unmöglich, auch nur annähernd jede Arbeit hier nennen zu wollen. Dem Fachmann sagen solche Namen wie Rice, Hutchinson, Yokobori, Mura, Weertman — um nur einige der klangvollsten anzuführen — bereits, daß auf dem Symposium ein kompetentes Gremium über aktuelle Schwerpunkte der Bruchforschung diskutiert hat. Die acht Sitzungen wurden ebenfalls von führenden Wissenschaftlern geleitet und diskutiert, wobei ein besonderer Wert der Proceedings auch in der abgedruckten Diskussion besteht, wie das bei bedeutenden IUTAM-Symposien zur guten Tradition gehört. Für den aktiven Forscher auf basagtem Gebiet sind die aussagefähigen Quellenverzeichnisse der meisten Autoren sehr nützlich, die Literatur bis 1980 enthalten. Man kann sagen, daß eigentlich alle wichtigen Fragen der Thematik des Symposiums in den Proceedings zumindest angesprochen werden, wobei einige auf Grund des begrenzten Umfanges recht kurz wegkommen mußten. Die Probleme reichen von der Mikrovoid-Koaleszenz bis zu den thermodynamischen konstitutiven Relationen und zu verallgemeinerten Integralkonzepten für allgemeineres Werkstoffverhalten im Rißspitzengebiet (NGUYEN, BUI, Frankreich). Es ist mit gutem Recht feststellbar, daß dieser Band ein wertvolles Werk für den Spezialisten der Bruchmechanik und darüber hinaus der Bruchforschung darstellt. Es ist aber auch für Kollegen gut lesbar geschrieben, die sich einen Überblick über aktuelle Fragen dieses Fachgebietes verschaffen möchten. Hervorzuheben ist noch die Zusammenstellung der Teilnehmerliste mit den Angaben der Forschungsinstitute der Anwesenden aus 15 Teilnehmerländern, womit gleichzeitig ein Verzeichnis eines gewichtigen Teils führender Schulen der Bruchmechanik gegeben wird.

Karl-Marx-Stadt B. MICHEL

Kreißig, R. / Drey, K.-D. / Naumann, J., Methoden der Plastizität. Anwendung auf Umformprobleme. München-Wien, Carl Hanser Verlag 1980. 204 S., 80 Abb., DM 32,–. ISBN 3-446-13148-5

Das Buch umfaßt drei jeweils in sich geschlossene Beiträge zur Anwendung der Plastizitätstheorie auf Probleme der Umformtechnik. Es handelt sich um einen Nachdruck der 1980 im VEB Fachbuchverlag Leipzig erschienenen Auflage.

R. KREISSIG untersucht die Berechnung stationärer Umformvorgänge mit Hilfe der Theorie des ebenen, starr-idealplastischen Deformationszustandes. Dargelegt werden die Grundlagen der Theorie, Lösungsmöglichkeiten kinematisch bestimmter Aufgaben am Beispiel des Fließpressens bzw. des Ziehens und abschließend die Verbesserung der Resultate durch Erfassung der Anisotropie, Inhomogenität und Verfestigung. K.-D. DREY entwickelt ein Verfahren zur Berechnung ebener viskoplastischer Umformprobleme auf der Grundlage der FEM, das von der Lösung eines zugeordneten starr-idealplastischen Werkstoffs ausgeht und die nichtlinearen Anteile des verwendeten erweiterten Stoffgesetzes des Bingham-Körpers iterativ berücksichtigt. Besonderer Wert wird auf die Erfassung des isotropen Druckes gelegt. J. NAUMANN nutzt das Moiréverfahren zur experimentell-theoretischen Analyse von Umformvorgängen. Nach der Beschreibung der Grundlagen des Moiréverfahrens wird die Versuchsauswertung ausführlich behandelt. Als Anwendungsbeispiele werden in erster Linie verschiedene Formen des Strangpressens betrachtet.

In den einzelnen Beiträgen wird die Absicht deutlich, praktisch nutzbare Verfahren vorzustellen. Das ist durch die entwickelten Konzepte für die rechentechnische Realisierung der theoretischen Methoden und die vermittelten Erfahrungen zur Versuchsmethodik und Versuchstechnik auch gelungen.

Berlin W. KAISER

Böhme, G., Strömungsmechanik nicht-newtonscher Fluide. LAMM 52. Stuttgart, B. G. Teubner 1981. 280 S., DM 34,–. ISBN 3-519-02354-7

In diesem Buch kommen wesentliche Teile der Rheologie aus der Sicht der angewandten Mechanik zur Darstellung. Es handelt sich um inkompressible laminare Einphasenströmungen, die, bis auf einige Ausnahmen, bei konstanter Temperatur erfolgen. Als theoretische Grundlage dienen die Begriffe der Kontinuumsmechanik. Neben den nichtlinearen Fließeigenschaften, durch welche sich die nicht-Newtonschen Medien von den gewöhnlichen zähen Flüssigkeiten unterscheiden, wird ausführlich auf Normalspannungseffekte und Gedächtniseinflüsse eingegangen.

Die Darstellung beginnt mit den kontinuumsmechanischen Grundlagen, welche etwa ein Viertel des Umfangs in Anspruch nehmen. In den folgenden Kapiteln wird die Theorie an Hand konkreter mechanischer Erscheinungen entwickelt. Auf die Behandlung derjenigen Stoffeigenschaften, die bei stationären Schichtenströmungen hervortreten, folgen Vorgänge, welche durch die Fließfunktion kontrolliert werden, und die Effekte, die auf Normalspannungsdifferenzen zurückzuführen sind, wie z. B. der Weißenberg Effekt und die Strangaufweitung. Bei der Behandlung einfacher instationärer Strömungen sind die grundsätzlichen Bemerkungen über die drei Stoffunktionale sehr knapp gehalten; dafür werden erfreulicherweise eine Reihe von Beispielen der linearen Viskoelastizität und nichtlineare Effekte bei der instationären Rohrströmung behandelt. Anschließend finden wir Betrachtungen über fastviskosimetrische Strömungen und Dehnströmungen.

Zur theoretischen Beschreibung gewisser markanter Strömungserscheinungen, welche in nichtnewtonschen Fluiden auftreten, in gewöhnlichen zähen Flüssigkeiten aber nicht beobachtet werden, dient der Begriff der Sekundärströmung, d. h. die tatsächliche Bewegung wird aufgefaßt als Überlagerung der Strömung einer zähen Flüssigkeit und einer zusätzlichen „Sekundärströmung", zu deren Berechnung Strömungsverfahren herangezogen werden. Mit dieser Methode werden u. a. behandelt die konische Düsenströmung, die Strömungen um rotierende Körper und durch gekrümmte Rohre sowie die periodische Rohrströmung.

Insgesamt handelt es sich um eine klare Darstellung der Strömungsmechanik nicht-newtonscher Flüssigkeiten; die Theorie wird soweit entwickelt, wie es die Erfassung der beobachteten Erscheinungen erfordert. Durch Aufgaben am Ende jeden Kapitels wird der Studierende zu aktiver Mitarbeit angeregt.

Berlin J. FÖRSTE

Rosemann, H., Zuverlässigkeit und Verfügbarkeit technischer Anlagen und Geräte. Mit praktischen Beispielen von Berechnung und Einsatz in Schwachstellenanalysen. Berlin-Heidelberg-New York, Springer-Verlag 1981. XII, 188 S., 96 Abb., DM 48,–. US $ 22.40. ISBN 3-540-11000-3

Je umfangreicher und komplizierter Maschinen und Anlagen sind, je mehr verschiedene Bauteile zur Gewährleistung der Funktion notwendig sind, desto größer werden Ausfallmöglichkeiten und Fehlerhäufigkeiten. Da sich Fehler in der Regel nicht völlig ausschließen lassen, bestehen wesentliche Forderungen der modernen Industrie darin, durch geeignete Methoden Aussagen und Abschätzungen über die Zuverlässigkeit und Verfügbarkeit zu erhalten und eine hinreichende Anzahl von Ingenieuren heranzubilden, die diese Aufgaben bearbeiten können. Insbesondere dem zweiten Anliegen ist die vorliegende Monographie gewidmet. Zur Einführung werden in breiter Form die Grundbegriffe der Wahrscheinlichkeitsrechnung erläutert. Daran schließt sich die Behandlung verschiedener Anordnungen — Reihenanordnung, Parallelanordnung, Kombination von Reihen- und Parallelanordnungen, r aus n Anordnung — an, wobei jeweils Formeln zur Berechnung der Zuverlässigkeit bzw. Ausfallwahrscheinlichkeit angegeben werden. Mit Hilfe dieser Grundanordnungen wird die Analyse beliebiger Anordnungen zumindest näherungsweise möglich. Das entsprechende Vorgehen ist in einer algorithmischen Form dargelegt und an Hand zahlreicher Beispiele illustriert. Die statischen Betrachtungen werden durch die Einführung der Zeitvariablen wesentlich erweitert, so daß Verteilungsfunktionen und -dichten der Ausfälle, Ausfallrate, Überlebenswahrscheinlichkeit, mittlere Zeit bis zum nächsten Ausfall u. ä. als neue Kenngrößen auftreten. Dem schließt sich die Behandlung verschiedener Modelle für Zuverlässigkeit und Verfügbarkeit an.

Mit Hilfe dieser theoretischen Grundlagen werden umfangreiche Schwachstellen- und Zuverlässigkeitsanalysen von praktischen Beispielen durchgeführt, so für Wälzlagersysteme, Schiffsantriebsanlagen und elektrische Schaltanlagen.

Im abschließenden Literaturverzeichnis werden einige grundlegende Monographien zu diesem Gebiet vermißt, so GNEDENKO — BELJAJEW — SOLOWJEW: Mathematische Methoden der Zuverlässigkeitstheorie I und II, Akademie-Verlag, Berlin 1968; KOSLOW — USCHAKOW: Handbuch zur Berechnung der Zuverlässigkeit, Akademie-Verlag, Berlin 1978.

Insgesamt wird das übersichtlich abgefaßte Buch, das sowohl theoretische Grundlagen als auch numerische Berechnungsverfahren erläutert, Interessenten, die in das Gebiet eindringen wollen, eine wertvolle Hilfe sein.

Berlin H. FRIEDRICH

Cremer, L., Physik der Geige. Stuttgart, S. Hirzel Verlag 1981. 368 S., DM 118,–. 167 Abb., ISBN 3-7776-0372-4

Der international bekannte Raumakustiker legt mit dieser Monographie ein Werk über die Physik der Streichinstrumente vor, wie es in deutscher Sprache nur wenige gibt. Vergleicht man die Darstellung mit der von C. V. Raman — derselbe, der 1930 den Nobelpreis für seine spektroskopischen Arbeiten erhielt —: Musikinstrumente und ihre Klänge [Geiger-Scheel, Handbuch der Physik, Bd. 8, S. 369—397 (1927)], dann wird die Weiterentwicklung in der Erforschung der Physik der Saiteninstrumente besonders augenfällig. Der Verf. stützt sich auf viele veröffentlichte und unveröffentlichte Diplom- und Doktorarbeiten, die unter seiner Leitung als Direktor des Instituts für technische Akustik an der Technischen Hochschule Berlin(West) angefertigt wurden. Der Leser findet in der Monographie — die sich ja nicht nur an Akustiker wendet — aber auch viel Altbekanntes. Völlig unberücksichtigt bleiben der Einfluß des Lacks und die Gütemerkmale alter Meistergeigen.

Das Buch gliedert sich in drei Teile: Der Teil I beschreibt den Streichvorgang (S. 21—169), der Teil II den Instrumentenkörper (S. 171—298) und der Teil III den abgestrahlten Schall (S. 299—363) einschließlich der Einflüsse des umgebenden Raumes.

Das Problem der durch den Streichvorgang hervorgerufenen selbsterregten Saitenschwingungen wird schrittweise behandelt. Über die Berücksichtigung der Energieverluste, der Torsion und der Biegesteife der Saite führt der Weg zu einer vollständigen Lösung des Problems. Das Kap. 8 (Teil I) dürfte für die Leser der ZAMM besonders von Interesse sein. Für die periodische Lösung ergibt sich eine Hammersteinsche Integralgleichung, deren Kern allerdings höchstens näherungsweise bekannt ist. Durch einen Iterationsprozeß ergeben sich hier Resultate, die übereinstimmen mit einer Simulation des Streichvorgangs im Computer. Die Integralgleichungsmethode gestattet eine Verallgemeinerung auf andere selbsterregte Systeme.

Der Teil II beginnt mit einer Untersuchung der Schwingungen des Steges und fährt fort mit Modellrechnungen für den Corpus als einer Kombination schwingender Kontinua. Dabei sagt der Verf.: „Eine Vorausberechnung, wie sie etwa der Flugzeugkonstrukteur bei seinen Kombinationen aus Stäben, Platten und Schalen unter Computereinsatz vornimmt, wird es nie geben. Sie ist aber auch nur dort unabdingbar, wo es um Menschenleben geht. Auf dem Gebiet der Kunst, wo mit Instrumenten ‚gespielt' wird, ist die Gestaltungsfreiheit größer. Hätte ein Ingenieurbüro, das mit allen heutigen Kenntnissen und Mitteln der Technik versehen ist, den Auftrag gehabt, ein Streichinstrument zu entwickeln, es hätte nie seine vorhandene, empirisch entwickelte Form gewonnen" (S. 238—239).

Im letzten Teil wird mit Hilfe der Punktstrahlersynthese und durch Messungen das Schallfeld von Streichinstrumenten für verschiedene Fälle der Größenverhältnisse Quellenausdehnung/Wellenlänge untersucht. Die Einflüsse des umgebenden Raumes gehören in die Raumakustik und werden nur soweit kurz behandelt, wie sie speziell für Streichinstrumente wichtig sind.

Das Buch wendet sich an die Akustiker, darüber hinaus ist es für alle interessierten Naturwissenschaftler eine genußreiche Lektüre, insbesondere, wenn es sich dabei auch um Musikausübende handelt (was ja bei Mathematikern und Physikern nicht selten ist). Der Verf. wünscht, daß es auch bei Geigenbauern und Musikern Beachtung findet, obwohl er kritisch schreibt: „Wenn die Liebe, die wir Akustiker den Musikern und ihrem Schaffen entgegenbringen, nur selten auf Gegenliebe hoffen kann, so liegt das weniger an anderen Inhalten der Ausbildung, als daran, daß die Künstler allen Bemühungen, mit naturwissenschaftlichen Kategorien . . . in ihren Aufgabenbereich einzudringen, mit Skepsis gegenüberstehen" (S. 6). Der Ref. meint, daß das Buch geeignet ist, wenigstens bei einigen Paragraphen diese Skepsis in den Hintergrund treten zu lassen.

Berlin D. Ullmann

Zypkin, Ja. S., Grundlagen der Theorie automatischer Systeme. Berlin, VEB Verlag Technik 1981. 484 S., 362 Abb., M 65,—. BN 5529637. (Übersetzung aus dem Russischen)

Wenn ein so bedeutender Autor wie Ja. S. Zypkin ein neues Werk veröffentlicht, darf man auf den Inhalt gespannt sein. Das vorliegende Buch „Grundlagen der Theorie automatischer Systeme" ist eine umfassende Einführung in die Methoden der Regelungstechnik mit zahlreichen Hinweisen auf klassische und moderne Anwendungsbeispiele. Es hat jedoch nichts mit der modernen Regelungstheorie gemeinsam, die einen hohen, für den Ingenieur zum Teil unverständlichen Entwicklungsstand erreicht hat. Zypkin ist es in hervorragender Weise gelungen, die gesamte heutige Breite regelungstechnischer Verfahren mit Hilfe der elementaren Mathematik anschaulich darzustellen.

Das Buch gliedert sich in drei Teile. Nach einer Einführung werden im besonderen zeitkontinuierliche und zeitdiskrete Systeme ausführlich behandelt, wobei jeweils lineare und nichtlineare Systeme unterschieden werden. Der erste Teil, die Einführung, besitzt rein beschreibenden Charakter. Er erleichtert dem unerfahrenen Leser die Einarbeitung.

Der zweite Teil beginnt mit der mathematischen Beschreibung linearer Systeme und klassifiziert die Regelkreise nach ihrer Struktur. Dann folgen eine Diskussion der technisch wichtigen Frage der Parameterempfindlichkeit, die Darstellung von Frequenzgang und Übertragungsfunktion sowie die Bestimmung der statischen Regelabweichung. Die Stabilitätsbetrachtung umfaßt die Definitionen und die Möglichkeiten zur Stabilisierung strukturell instabiler Systeme. Die Übergangsfunktion führt auf die Gütekriterien, wobei die quadratische Regelfläche im einzelnen berechnet wird. Weiterhin werden zufällige Erregungen betrachtet. Für nichtlineare Systeme werden die Stabilitätsdefinitionen erweitert und die absolute Stabilität mit Hilfe der Frequenzganggleichungen überzeugend dargestellt. Die harmonische Linearisierung periodisch erregter Systeme und die statistische Linearisierung zufällig erregter Systeme schließen den zweiten Teil ab.

Im dritten Teil werden zunächst Fragen des Abtastvorgangs und der Diskretisierung erläutert. Dann folgen die linearen zeitdiskreten Systeme, wobei wiederum alle wesentlichen Punkte abgehandelt werden: Regelkreisstrukturen, Übertragungsfunktion, Regelabweichung, Stabilität, Gütekriterien und Zufallserregung. Zusätzlich werden die für zeitdiskrete Impulssysteme charakteristischen Syntheseaufgaben der endlichen Einstellzeit und der Kompensation der Totzeit beschrieben. Die nichtlinearen Systeme erfordern eine erneute Diskussion der mathematischen Beschreibung, der absoluten Stabilität und der periodischen Erregung. Schließlich wird noch der Übergang zwischen den diskreten und kontinuierlichen Systemen und eine verschwindende Abtastzeit aufgezeigt. Der Anhang enthält einige kurze Beiträge zur Laplace- und Fouriertransformation. Das Literaturverzeichnis umfaßt hauptsächlich Lehrbücher zu den einzelnen Themen.

Das vorliegende Werk kann allen Regelungstechnikern ohne Einschränkung empfohlen werden. Der klassisch ausgebildete Ingenieur findet viele praxisrelevante Verfahren, und der mit modernen Methoden vertraute Wissenschaftler wird darüber erstaunt sein, was mit elementaren Mitteln alles möglich ist. Trotzdem hätte sich der Referent wenigstens einen kurzen Ausblick auf die Methoden im Zustandsraum gewünscht, die auch das große Gebiet der gekoppelten Mehrfachregelsysteme erschließen.

Stuttgart W. Schiehlen

Kevorkian, J. / Cole, J. D., Perturbation Methods in Applied Mathematics. Applied Mathematical Sciences 34. Berlin-Heidelberg-New York, Springer-Verlag 1981. X, 558 S., 79 Abb., DM 88,—. US $ 41.90. ISBN 3-540-90507-3

Das vorliegende Buch stellt eine überarbeitete und ergänzte Fassung einer 1968 unter demselben Titel erschienenen Monografie von J. D. Cole dar. Das wesentliche Ziel des Buches besteht darin, Näherungsmethoden zur Untersuchung singulär gestörter Probleme für gewöhnliche und partielle Differentialgleichungen darzulegen, wobei die asymptotischen Entwicklungen im Mittelpunkt stehen. Dabei ist es das ausgesprochene Ziel der Verfasser, keine mathematische Theorie solcher Näherungsmethoden anzugeben (Konvergenzfragen, Existenz einer exakten Lösung, Fehlerabschätzungen werden nicht erörtert), vielmehr zeigen sie anhand verschiedener und im allgemeinen vereinfachter Aufgabenstellungen, wie man durch im allgemeinen physikalisch motivierte Überlegungen zur Aufstellung eines geeigneten Näherungsverfahrens kommt.

Kapitel 1 enthält Grundkenntnisse aus der Theorie asymptotischer Entwicklungen. In Kapitel 2 wird die Methode der Grenzprozeßentwicklung für gewöhnliche Differentialgleichungen beschrieben. Im allgemeinen handelt es sich um Gleichungen zweiter Ordnung, wobei die höchste Ableitung mit einem kleinen Faktor versehen ist. Kapitel 3 ist den kumulativen Störungsproblemen gewidmet. Die asymptotischen Entwicklungen werden unter Verwendung mehrerer Zeitskalen konstruiert. Es enthält u. a. Anwendungen auf Satellitenprobleme. Kapitel 4 beschäftigt sich mit der Behandlung singulär gestörter Probleme bei partiellen Differentialgleichungen zweiter Ordnung mittels der Grenzprozeßentwicklung, wobei speziell Grenzschichtprobleme viskoser inkompressibler Medien erörtert werden. Kapitel 5 enthält typische Anwendungsbeispiele für asymptotische Entwicklungen aus der Hydromechanik.

Nach jedem Abschnitt werden weiterführende Literaturzitate angegeben. Leider gibt es nur wenige Hinweise auf tiefergehende mathematische Literatur; die grundlegenden Untersuchungen von VIŠIK sowie die Arbeiten der Moskauer Schule um TICHONOV und VASILJEVA werden z. B. überhaupt nicht erwähnt.

Insgesamt gesehen kann die Monografie als eine gute Einführung in die Methode der asymptotischen Entwicklungen zur Behandlung singulär gestörter Probleme bezeichnet werden.

Berlin K. R. SCHNEIDER

Carr, J., Applications of Centre Manifold Theory. Applied Mathematical Sciences 35. Berlin-Heidelberg-New York, Springer-Verlag 1981. XII, 142 S., DM 32,—. US $ 14.60. ISBN 3-540-90577-4

Eine wesentliche Vereinfachung der qualitativen Untersuchung autonomer Differentialgleichungssysteme in der Umgebung von Gleichgewichtslagen kann im allgemeinen durch die Verwendung der sogenannten Zentrumsmannigfaltigkeiten erreicht werden. In Analogie zur Reduktionsmethode von LJAPUNOV-SCHMIDT bei der Untersuchung von Bifurkationsproblemen gestatten die Zentrumsmannigfaltigkeiten im allgemeinen eine wesentliche Erniedrigung der Ordnung des zu betrachtenden Differentialgleichungssystems.

In der vorliegenden Monografie werden Systeme gewöhnlicher Differentialgleichungen der Gestalt

$$dx/dt = Ax + f(x, y) , \qquad dy/dt = By + g(x, y) , \qquad (1)$$

wobei A eine stabile Matrix ist, während B nur Eigenwerte mit verschwindendem Realteil besitzt und Differentialgleichungen der Form $dw/dt = Cw + N(w)$ in einem unendlich dimensionalen Banachraum betrachtet, wobei C den infinitesimalen Erzeuger einer stark stetigen Halbgruppe darstellt. Die Funktionen f, g und N verschwinden zusammen mit ihrer ersten Ableitung im Ursprung. Eine Integralmannigfaltigkeit von (1) mit der Darstellung $y = h(x)$, die für $|x| < \delta$ definiert ist und den Relationen $h(0) = 0$, $h'(0) = 0$ genügt, wird als Zentrumsmannigfaltigkeit von (1) bezeichnet. Es werden drei fundamentale Sätze über Zentrumsmannigfaltigkeiten (Existenzsatz, Vergleichssatz, Approximationssatz) formuliert und deren Beweis skizziert.

Ein wesentliches Anwendungsgebiet der Theorie der Zentrumsmannigfaltigkeiten ist die Bifurkationstheorie dynamischer Systeme. In der vorliegenden Monografie wird die Hopfbifurkation autonomer Systeme im generischen Fall und für singulär gestörte Systeme betrachtet. Als Beispiele für die Anwendung von Zentrumsmannigfaltigkeiten auf Differentialgleichungen im Banachraum werden u. a. parameterabhängige Systeme semilinearer parabolischer und hyperbolischer Gleichungen betrachtet. Eine ausführliche Diskussion auf knapp einem Viertel der Monografie eines zweidimensionalen autonomen Systems, das im Ursprung zwei Nullwurzeln besitzt und von zwei Parametern abhängt, ist vom qualitativen Gesichtspunkt recht interessant, hängt aber nicht unmittelbar mit den Zentrumsmannigfaltigkeiten zusammen.

Die vorliegende Monografie stellt eine leicht verständliche Einführung in die Theorie der Zentrumsmannigfaltigkeiten und ihrer Anwendungen dar.

Berlin K. R. SCHNEIDER

Bohl, E., Finite Modelle gewöhnlicher Randwertaufgaben. LAMM 51. Stuttgart, B. G. Teubner 1981. 318 S., DM 29,80. ISBN 3-519-02353-9

Dargestellt und illustriert werden Theorie und Anwendung numerischer Verfahren zur Lösung von Randwertproblemen gewöhnlicher Differentialgleichungen zweiter Ordnung. Dabei werden an Systemen nur die betrachtet (und darin liegt vielleicht ein Risiko des Buches), die sich auf eine Gleichung zweiter Ordnung zurückführen lassen. Die Darstellung ist gründlich, leicht lesbar und interessant durch die zahlreichen Beispiele aus verschiedenen Anwendungsgebieten (Stabbiegung, chemische und biochemische Reaktionen). An numerischen Methoden werden die Diskretisierung (auch höherer Ordnung) über endliche Differenzen oder finite Elemente und, zur Lösung der entstandenen Gleichungssysteme, die einfache (Picardsche) Iteration, das Parallelen- und das Newtonverfahren und Fortsetzungsverfahren besprochen. Viel Aufmerksamkeit wird den bei angewandten Aufgaben häufig auftretenden Schwierigkeiten gewidmet, die verursacht sind durch mehrfache Lösbarkeit (Hysteresis, Verzweigung, stabile und instabile Lösungen) und durch kleine Parameter (Grenzschichten).

Das Buch regt zum Experimentieren an und eignet sich durch die vielen numerischen Beispiele auch zum Testen entsprechender Programmpakete.

Berlin G. STOYAN

Crouch, T., Matrix Methods Applied to Engineering Rigid Body Mechanics. Oxford-New York-Toronto-Sydney-Paris-Frankfurt, Pergamon Press 1981. XVIII, 339 S., DM 43,—. ISBN 0-08-024246-4

Wie der Titel ausdrückt, ist das vorliegende Lehrbuch ganz auf die Darlegung und Anwendung von Matrizenmethoden für eine Ingenieur-Starrkörpermechanik gerichtet. Die Gliederung umfaßt Kinematik, Dynamik, Anwendungsbeispiele für kinematische Probleme und Anwendungsbeispiele für dynamische Probleme. Der Stoff ist so verteilt, daß auf die ersten beiden Abschnitte nur 64 Seiten entfallen, ein knappes Fünftel des gesamten Umfanges. Es werden sehr wichtige Gebiete der Starrkörperdynamik, wie Fragen der Stabilität von Bewegungen, Kreiselbewegungen und Mechanismendynamik nur anhand der 65 mit ausführlichen Lösungen dargelegten Beispiele behandelt. Weitere 72 Aufgaben, deren Lösungen angegeben sind, ermöglichen dem interessierten Leser die Einarbeitung in ein anspruchsvolles und interessantes Gebiet der Mechanik. Hervorgehoben werden muß die außerordentliche Sorgfalt, mit der die Kinematik dargelegt wird. Eine spezielle Notation sichert zum Beispiel, daß bei keiner Gleichung Unklarheiten über das Bezugssystem eines Vektors bestehen. Dagegen erhebt der Referent Bedenken dagegen, daß in der Dynamik die Kraft nur als Produkt von Masse und Beschleunigung eingeführt wird. Die für die theoretische Mechanik so wichtige Unterscheidung zwischen eingeprägten und Zwangskräften wird nicht erwähnt, auch fehlt jeder Hinweis auf das D'ALEMBERTsche Prinzip, die Lagrangeschen Bewegungsgleichungen oder Variationsprinzipien. Zugestehen muß man jedoch, daß der Autor auch relativ komplizierte Probleme der Starrkörpermechanik ohne diese Theoreme zu lösen versteht und den Lösungsweg, unterstützt durch klar gestaltete Zeichnungen, in gut verständlicher Form vorführt. Insgesamt und besonders wegen der vielen instruktiven Beispiele ist ein Buch entstanden, das nicht nur dem Ingenieurstudenten, sondern auch dem Praktiker, der vor Aufgaben dieser Art gestellt ist, eine wertvolle Hilfe sein kann.

Magdeburg U. FISCHER

Huang, T. S. (ed.), Image Sequence Analysis. Springer Series in Information Sciences 5. Berlin-Heidelberg-New York, Springer-Verlag 1981. XVI, 437 S., 125 Abb., DM 85,—. US $ 39.60. ISBN 3-540-10919-6

Das in dem Sammelband behandelte Gebiet ist sicher einer der schwierigsten und neuesten der Signalverarbeitung. Sieht man von einigen naheliegenden Grundansätzen ab (Vergleich der räumlichen und zeitlichen Änderungen, Bestimmung der Phasenverschiebung in der räumlichen Fouriertransformation), so scheint wenigstens bisher kaum eine durchgehende Theorie zu existieren. Es wird vielmehr versucht, sich mit heuristischen Ansätzen den Problemen anzupassen und sie damit zu einer halbwegs praktikablen Lösung zu führen.

Den Leser, der die relative Strenge der älteren Signaltheorie gewohnt ist, wird das Buch deshalb zunächst wohl enttäuschen. Es kann aber als verdienstvoll gelten, daß keiner der Autoren versucht, axiomatische Strenge vorzutäuschen, wo offenbar nur pragmatische Ansätze weiterführten. Besonders deutlich wird dieser Trend im Kapitel 2 „Was kann aus der Anwendung gelernt werden?“. Das erklärte Ziel des Verfassers ist es, hier eine große Fülle von Einzelbeispielen zu bringen, die nur soweit aufgearbeitet werden, daß der Leser Querverbindungen zu seiner Aufgabe erkennen kann und damit eine Auswahl aus der sehr reichlich angebotenen Literatur möglich ist, Informativ ist auch die Gegenüberstellung der Kapitel 4 „Image sequence enthancement"; „Image region extraction of movingobjects" und 7 „processing of medical image sequences", bei denen auf den ersten Blick sehr ähnliche Probleme behandelt werden. Während aber für die Signal-Rausch-Verbesserung durch rekursive Filterung längs der geschätzten Bewegungstrajektorie der Bildelemente bereits eine relativ anspruchslose Geschwindigkeitsschätzung nach der Gradientenmethode genügt, und bei einer medizinischen Anwendung (Verfolgung der Ventrikel-Bewegung mittels einer Röntgenbildfolge) die Bewegung selbst wesentlich zur Identifizierung des Objektes genutzt werden kann, erfordert die Verfolgung bewegter Objekte in Verkehrsszenen anscheinend eine aufwendige Identifizierung des Objektes durch Transformation von Pixel-Gruppen in neue Merkmalsräume und Suche nach konsistenten Gebieten. In dem Buch werden außerdem eine Reihe von Beiträgen über Datenkompression von Fernsehbildern zum Zwecke der Übertragung auf relativ schmalbandigen Strecken bzw. der Filterung dargeboten. Bei diesen Fällen spielt wiederum die Psychologie des Sehvorganges eine bedeutende Rolle. An diesem und einer Reihe anderen Beispielen wird meiner Meinung nach überzeugend demonstriert, wie stark die Methoden von Details der Aufgabenstellung abhängig sind. Aussagen darüber, inwieweit diese Verfahren bereits in Echtzeit realisierbar sind, lassen sich nur sehr spärlich finden. Lediglich im Kapitel 5 waren detailliertere Angaben hierzu zu finden, aus denen hervorgeht, daß die Bearbeitungszeit noch um ein bis zwei Größenordnungen über der Echtzeit liegt.

Das Buch ist anscheinend in erster Linie für einen Leserkreis gedacht, der sich intensiver mit der Thematik beschäftigt und dann auch entsprechenden Nutzen aus den sehr umfangreichen Literaturhinweisen und der Verfolgung der hier gegebenen Anregungen ziehen kann.

Berlin K.-H. Schmelovski

In der DDR beziehbar im Rahmen des Kontingents.

Heidmann, J., Relativistic Cosmology. An Introduction. Berlin-Heidelberg-New York, Springer-Verlag 1980. XV, 168 S., 45 Abb., DM 48,—. US $ 28.30. ISBN 3-540-10138-1

Dieses Buch ist Abd-al-rahman gewidmet, der im Jahre 964 auf einer Sternkarte den Fleck einzeichnete, der später Andromeda-Nebel genannt wurde, und dessen wahre Natur erst 1923 von Edwin Hubble erkannt wurde, womit die moderne (auf Naturwissenschaft, nicht mehr allein auf philosophische Spekulation gegründete) Kosmologie beginnt. Das Buch ist über weite Passagen in Form eines Berichtes gehalten und bringt so das Kunststück fertig, auch für Leser, deren Vorbildung zum vollen Verständnis des gesamten Textes nicht ausreichen sollte, bis zum Schluß interessant zu bleiben. Kap. 1 (The Metagalaxy out to a Distance of one Gigaparsec) gibt einen Überblick über das Beobachtungsmaterial und das daraus erschlossene Bild von der Struktur des Universums im „Nahbereich", wo relativistische Effekte vernachlässigbar sind. Kap. 2 (Spaces with Constant Curvature) bringt rein geometrische Betrachtungen, Kap. 3 (Model Universes) verbindet diese mit physikalischen Aspekten zu relativistischen Modellen. Darauf aufbauend wird mit dem vierten Kapitel (The Metagalaxy in the Relativistic Zone) das Titelthema erreicht und das Beobachtungsmaterial für die äußere Metagalaxis interpretiert.

Rostock M. Heinrich

EINGEGANGENE BÜCHER

Bei der Schriftleitung sind folgende Bücher eingegangen (ausführliche Besprechung bleibt vorbehalten).

Engeler, E. (ed.), Logic of Programs. Proceedings, Zürich 1979. Lecture Notes in Computer Science 125. Berlin-Heidelberg-New York, Springer-Verlag 1981. V, 245 S., DM 25,—. US $ 11.70. ISBN 3-540-11160-3

Thirring, W., Quantum Mechanics of Atoms and Molecules. A Course in Mathematical Physics 3. Wien-New York, Springer-Verlag 1981. VIII, 300 S., s 465,—. DM 65,—. US $ 27.70. ISBN 3-211-81620-8

Bourbaki, N., Eléments de mathématique. Algèbre. Chapitres 4 à 7. Paris, MASSON 1981. 422 S., F 290.—. ISBN 2-225-68574-6

Arnold, L. / Lefever, R. (eds.), Stochastic Nonlinear Systems in Physics, Chemistry, and Biology. Proceedings, Bielefeld, FRG, 1980. Berlin-Heidelberg-New York, Springer-Verlag 1981. VIII, 237 S., 48 Abb., DM 62,—. US $ 29.60. ISBN 3-540-10713-4

Ostrowsky, Yu. I. / Butusov, M. M. / Ostrovskaya, G. V., Interferometry by Holography. Springer Series in Optical Sciences 20. Berlin-Heidelberg-New York, Springer-Verlag 1980. X, 330 S., 184 Abb., DM 64,—. US $ 37.80. ISBN 3-540-09886-0

NACHRICHTEN

24th Polish Solid Mechanics Conference

The Conference will be held on the 22nd—27th August 1983 in or near Warsaw. The Conference will cover all major areas of the mechanics of solids and structures with particular emphasis on the connections between continuum and micro-structural propeties of materials and the solution of practically significant problems.

The organizing Committee invites the authors to submit the papers for the technical sessions. Interested authors are kindly requested to send a one page abstract preferably in English by the 1—st of March 1983.

SECRETARY: Krzysztof Podolak

CHAIRMAN: Bogdan Raniecki

Address for correspondence: Institute of Fundamental Technological Research, Polish Academy of Sciences, Świętokrzyska 21, 00-049 Warsaw, Poland, Telephone 268802, Telex 815638 ippt pl.

FSC
www.fsc.org
MIX
Papier | Fördert gute Waldnutzung
FSC® C083411

Kolibri 360 GmbH
Ferdinand-Jühlke-Straße 7
99095 Erfurt, Deutschland
produktsicherheit@kolibri360.de